Carbon Nanotube-Based Nanocomposites

Carbon Nanotube-Based Nanocomposites

Editor

Anna Boczkowska

MDPI • Basel • Beijing • Wuhan • Barcelona • Belgrade • Manchester • Tokyo • Cluj • Tianjin

Editor
Anna Boczkowska
Warsaw University of Technology
Poland

Editorial Office
MDPI
St. Alban-Anlage 66
4052 Basel, Switzerland

This is a reprint of articles from the Special Issue published online in the open access journal *Materials* (ISSN 1996-1944) (available at: https://www.mdpi.com/journal/materials/special_issues/carb_nano_nanocompos).

For citation purposes, cite each article independently as indicated on the article page online and as indicated below:

LastName, A.A.; LastName, B.B.; LastName, C.C. Article Title. *Journal Name* **Year**, *Volume Number*, Page Range.

ISBN 978-3-0365-2201-2 (Hbk)
ISBN 978-3-0365-2202-9 (PDF)

Cover image courtesy of Anna Boczkowska, the Guest Editor

© 2021 by the authors. Articles in this book are Open Access and distributed under the Creative Commons Attribution (CC BY) license, which allows users to download, copy and build upon published articles, as long as the author and publisher are properly credited, which ensures maximum dissemination and a wider impact of our publications.

The book as a whole is distributed by MDPI under the terms and conditions of the Creative Commons license CC BY-NC-ND.

Contents

About the Editor vii

Preface to "Carbon Nanotube-Based Nanocomposites" ix

Mateusz Mucha, Aneta Krzyzak, Ewelina Kosicka, Emerson Coy, Mikołaj Kościński, Tomasz Sterzyński and Michał Sałaciński
Effect of MWCNTs on Wear Behavior of Epoxy Resin for Aircraft Applications
Reprinted from: *Materials* 2020, 13, 2696, doi:10.3390/ma13122696 1

Vanessa Vilela Rocha and Péter Ludvig
Influence of Carbon Nanotubes on the Mechanical Behavior and Porosity of Cement Pastes Prepared by A Dispersion on Cement Particles in Isopropanol Suspension
Reprinted from: *Materials* 2020, 13, 3164, doi:10.3390/ma13143164 19

Agata Zubkiewicz, Anna Szymczyk, Piotr Franciszczak, Agnieszka Kochmanska, Izabela Janowska and Sandra Paszkiewicz
Comparing Multi-Walled Carbon Nanotubes and Halloysite Nanotubes as Reinforcements in EVA Nanocomposites
Reprinted from: *Materials* 2020, 13, 3809, doi:10.3390/ma13173809 37

Paulina Nakonieczna-Dabrowska, Rafał Wróblewski, Magdalena Płocińska and Marcin Leonowicz
Impact of the Carbon Nanofillers Addition on Rheology and Absorption Ability of Composite Shear Thickening Fluids
Reprinted from: *Materials* 2020, 13, 3870, doi:10.3390/ma13173870 61

Paulina Latko-Durałek, Rafał Kozera, Jan Macutkevič, Kamil Dydek and Anna Boczkowska
Relationship between Viscosity, Microstructure and Electrical Conductivity in Copolyamide Hot Melt Adhesives Containing Carbon Nanotubes
Reprinted from: *Materials* 2020, 13, 4469, doi:10.3390/ma13204469 71

Paweł Smoleń, Tomasz Czujko, Zenon Komorek, Dominik Grochala, Anna Rutkowska and Małgorzata Osiewicz-Powezka
Mechanical and Electrical Properties of Epoxy Composites Modified by Functionalized Multiwalled Carbon Nanotubes
Reprinted from: *Materials* 2021, 14, 3325, doi:10.3390/ma14123325 89

Minoj Gnanaseelan, Kristin Trommer, Maik Gude, Rafal Stanik, Bartlomiej Przybyszewski, Rafal Kozera and Anna Boczkowska
Effect of Strain on Heating Characteristics of Silicone/CNT Composites
Reprinted from: *Materials* 2021, 14, 4528, doi:10.3390/ma14164528 105

About the Editor

Anna Boczkowska DSc., PhD., Eng. is Vice Dean for Research of Faculty of Materials Science and Engineering, Warsaw University of Technology. She graduated from the Faculty of Materials Science and Engineering in 1989. She received her PhD in 2000 and DSc in 2011. She has been a Professor of Technical Sciences since 2018. She has authored over 200 scientific papers, books and patents. Since 2001, she has been the leader of many research and targeted projects, namely the scientific leader of Technology Partners teams in the ELECTROPOL, ELECTRICAL, SARISTU and PLATFORM FP7 / H2020 projects, and areas of research cover composites, polymers, nanomaterials, intelligent materials. She is the member of many scientific councils, national and international scientific societies, e.g., The Council of the Faculty of Materials Science and Engineering at the Warsaw University of Technology, Scientific Council of the Institute of Aviation, KOMPOZYT-EXPO trade fair, Materials Science and Metallurgy Committee of the Polish Academy of Sciences, Member of the Polish Carbon Society (PTW), Polish Society of Composite Materials (PTMK), International Society of Optical Engineering (SPIE), American Association for the Advancement of Science (AAAS), American Chemical Society (ACS).

Preface to "Carbon Nanotube-Based Nanocomposites"

Carbon nanotubes (CNTs) have been widely studied for years, due to their outstanding properties, i.e., high mechanical strength, thermal and electrical conductivity, and resistance to high temperature. The combination of the abovementioned properties, together with CNTs' low weight and high aspect ratio, makes them promising candidates as reinforcements for polymer, metal, or ceramic composites. In order to achieve the full potential of CNTs, two critical issues have to be solved: i) the homogeneous dispersion of CNTs in the matrix, ii) the interfacial bonding between the CNTs and the matrix. The most commonly studied CNTs are doped nanocomposites based on the polymer matrix. However, recently, there have also been extensive studies on the introduction of CNTs to the metal or ceramic matrix. The main problem in the application of CNT-based nanocomposites is most often related to the material properties, filtration and re-agglomeration of nanofiller, difficult processability, and in many cases, safety regulations. Of particular interest are recent developments in advanced composites, processes, characterization, and design. The potential market applications of CNT-based nanocomposites include automotive, aerospace, energy storage, and fuel cells.

The Special Issue "Carbon Nanotube-based Nanocomposites" (https://www.mdpi.com/journal/materials/special_issues/carb_nano_nanocompos) addresses advances in materials science, processing, characterization, technology development, and the application of nanocomposites based on carbon nanotubes.

Anna Boczkowska
Editor

Article

Effect of MWCNTs on Wear Behavior of Epoxy Resin for Aircraft Applications

Mateusz Mucha [1,*], Aneta Krzyzak [1], Ewelina Kosicka [2], Emerson Coy [3], Mikołaj Kościński [3,4], Tomasz Sterzyński [5] and Michał Sałaciński [6]

1. Faculty of Aviation, Military University of Aviation, Dywizjonu 303 35, 08-521 Dęblin, Poland; a.krzyzak@law.mil.pl
2. Faculty of Mechanical Engineering, Lublin University of Technology, Nadbystrzycka 36, 20-618 Lublin, Poland; e.kosicka@pollub.pl
3. NanoBioMedical Centre, Adam Mickiewicz University, Wszechnicy Piastowskiej 3, 61-614 Poznań, Poland; coyeme@amu.edu.pl (E.C.); mikolaj.koscinski@amu.edu.pl (M.K.)
4. Department of Physics and Biophysics, Faculty of Food Science and Nutrition, Poznań University of Life Sciences, Wojska Polskiego 38/42, 60-637 Poznań, Poland
5. Polymer Division, Institute of Materials Technology, Poznan University of Technology, Piotrowo 3, 61-138 Poznań, Poland; tomasz.sterzynski@put.poznan.pl
6. Air Force Institute of Technology, Księcia Bolesława 6, 01-494 Warsaw, Poland; michal.salacinski@itwl.pl
* Correspondence: m.mucha@law.mil.pl

Received: 1 May 2020; Accepted: 9 June 2020; Published: 12 June 2020

Abstract: The aim of the study is to assess the effect of multi-walled carbon nanotubes (MWCNTs) on the wear behavior of MWCNT-doped epoxy resin. In this study, a laminating resin system designed to meet the standards for motor planes was modified with MWCNTs at mass fractions from 0.0 wt.% to 2.0 wt.%. The properties of the carbon nanotubes were determined in Raman spectroscopy and HR-TEM. An examination of wear behavior was conducted on a linear abraser with a visual inspection on an optical microscope and SEM imaging, mass loss measurement, and evaluation of the wear volume on a profilometer. Moreover, the mechanical properties of MWCNTs/epoxy nanocomposite were evaluated through a tensile test and Shore D hardness test. The study shows that the best wear resistance is achieved for the mass percentage between 0.25 wt.% and 0.5 wt.%. For the same range, the tensile strength reaches the highest values and the hardness the lowest values. Together with surface imaging and a topography analysis, this allowed describing the wear behavior in the friction node and the importance of the properties of the epoxy nanocomposite.

Keywords: composites; carbon nanotubes; wear; mechanical properties; epoxy resin

1. Introduction

The observed development of science and technology leads to the use in machines and devices of such friction nodes in which the increased maximum load and speed of their movement occur. Due to the need to minimize the wear of tribological pairs [1,2], the possibilities of limiting the effects of friction are being investigated, which in effect is to contribute to the reduction of energy consumption and cooperating surfaces of machinery and equipment [3]. This is of vital importance, especially in the context of sensitive areas of material use, in which safety is taken into account before the economic criteria, which includes e.g., aviation industry [4,5]. Therefore, material engineering is becoming one of the key determinants of technological progress [6].

Currently, various modifiers allow achieving a wide range of properties for new materials [7–12]. As an example, studies conducted for aviation over modified polymers are most often carried out in order to improve material strength or electrical conductivity [13–15]. The improved electrical conductivity is desirable due to the lightning strike protection (LPS). In this case, it is important to determine the filler concentrations to obtain the best results. For instance in case of conductive carbon black in composites based on low-density polyethylene (LDPE) and poly(ethylene-co-vinyl acetate) (EVA), the percolation threshold was obtained for values higher than 15 wt.% [16].

The influence of MWCNTs on hardness, impact strength, and thermal conductivity of epoxy resin composites is widely discussed [10]. Other tests assessing the structural usefulness of a material include a hardness test [17,18]. Longitudinal and transverse shrinkage as well as plastic and elastic deformations [19] are also important in the context of materials for aviation [20,21]. Another important challenge in all technical applications of epoxy resin based composites is the need to produce materials with strongly reduced flammability [22,23].

Developing nanotechnology allows the use of increasingly sophisticated nanocomposites [24]. The nanofillers used include nanofibers [25,26], graphene flakes [27], and carbon nanotubes [28,29]. When descending to the nanoscale, dispersion of the nanofiller becomes crucial. The problems of inhomogeneous dispersion of nanofillers, agglomeration of MWCNTs, poor interfacial adhesion, etc. are widely discussed [30]. The creation of polymer nanocomposites with a homogeneous distribution of MWCNTs as a nanofiller is a very complex problem as shown in many papers [31–35]. To solve the dispersion problem, many solutions have been suggested, such as the addition of dispersing agents, shear mixing [28], and functionalization [29,36,37]. For example, the study on composites based on polypropylene (PP) presented in [38] suggests that the form of nanofiller dispersion such as the Masterbatch (MB) dilution is suitable for the dispersion of MWCNTs but at the expense of carbon nanotube shortening and, as a consequence, an increase in electrical resistivity. The homogeneity of MWCNT in the composite can be determined by means of the high-resolution scanning electron microscopy (HRSEM) [39].

The addition of nanoparticles to thermosetting matrix materials assists in strengthening the surface, which results in enhancing the tribological behavior of polymers. This is especially true for adhesive wear loading conditions under dry contact [40]. The filler addition greatly enhances the tribological properties of the epoxy resin, by reducing the friction coefficient and the wear rate [29]. As far as wear tests are concerned, a different test equipment is in use: from normalized ball on disc tribotesters to an in-house designed and built rubber wheel/dry sand test equipment [41] or original equipment designed and fabricated by the university [42].

In studying the application of carbon nanotubes to a piston ring and cylinder liner system, it was found that the friction coefficient decreases with an increase in MWCNTs content. That is a result of the fact that, under dry friction, the MWCNTs act as a solid lubricant and form a carbon film covering the contact surfaces [43]. For example, the study on the helical carbon nanotubes (H–CNTs) in the mass fraction range from 0.0 wt.% to 2.0 wt.%, presented in [44], showed that the friction coefficient, as well as the wear rate decrease with increasing the nanofiller content. Moreover, fillers can alter the crosslinking process of the polymeric matrix in comparison with the neat epoxy, in particular reduce the gel time of the resin [45,46].

Tools, such as TEM, are needed to verify the quality of nanocomposites. High-resolution transmission electron microscopy (HR-TEM) provides better insights into the physical behavior of many nanostructured materials [47]. An example of using it for research in which carbon nanotubes were used is presented in [48]. The image is formed by the interference of the diffracted beams with the direct beam. This is called phase contrast. HR-TEM images are obtained when the point resolution of the microscope is sufficiently high and a crystalline sample oriented along a zone axis [49].

Raman spectroscopy is also an important tool in the field of nanotechnology. Raman scattering is a component of Raman spectroscopic techniques and is used to obtain information about the structure and properties of molecules. This information comes from their vibrational frequencies. Raman scattering is a two-photon event. As known, the process focuses on the change in polarizability of a molecule which take account of its vibrational motion [50,51]. In [52], the authors focus on this method and describe its essence in detail. Some current applications of Raman spectroscopy are observed in the fields of biomedical diagnostics [53,54], archaeological science [55], industrial process control, environmental science [56], astrobiology, and materials engineering [57].

The investigated MWCNTs mass fraction range can vary depending on the aim of the research. In order to examine the impact on the electrical conductivity the wider range of 0.0–5.0 wt.% was applied in [10]. As far as mechanical properties are concerned, lower concentrations were investigated: 0.0–1.0 wt.% in [58], 1.0 wt.% in [59], 0.1–0.9 wt.% in [15], 0.0–0.5 wt.% in [42,60,61] (and also in [62] but the paper also cites studies for 5.0 wt.%), 0.3 wt.% in [63] (for of carboxyl functionalized MWCNTs), 0.025–0.2 wt.% in [46] and 0.2 wt.% in [37].

In the previous studies on laminating resin system designed to meet the standards for motor planes, modified with MWCNTs, proved that electrical conductivity of MWCNTs/epoxy nanocomposite increased with higher MWCNTs mass fraction even for a very simplified manufacturing procedure [64]. The percolation threshold was obtained in the range between 5.0 wt.% and 6.0 wt.%. However, such high mass fractions are not justified economically and due to the deterioration of mechanical properties and manufacturing problems (increased resin viscosity), are not recommended for structural components. Moreover, it is desirable that such parts exhibit good mechanical properties, such as high tensile strength and high resistance to wear [65,66]. In order to achieve a slight improvement of electrical properties and simultaneously provide a hope for an improvement of abrasion resistance, mass fractions from 0.0 wt.% to 2.0 wt.% were chosen.

The aim of the study is to assess the effect of MWCNTs content on the wear behavior of MWCNT-doped epoxy resin. The novelty of this article is that a linear abrader was used, which allows the comparison of results with measurements made by other abrasion tests. Secondly, Raman spectroscopy and TEM imaging were performed, which is not common in abrasive wear testing. Moreover, carbon nanotubes concentrations of up to 2.0% were used, which is not common in the studies of mechanical properties of epoxy resin nanocomposites. The presented results can be a reference point for the research on wear behavior, as well as epoxy resin modification in order to improve mechanical properties or electrical conductivity.

2. Materials

2.1. Raw Materials

The matrix of the MWCNTs/epoxy nanocomposite was made of the MGS L285 laminating resin system approved by German Federal Aviation Authority. L285 is intended for use with glass, carbon and aramid fibers and is characterized by high static and dynamic strength. After heating, it meets the standards for motor aircraft, gliders and powered sailplanes. The laminating resin L285 is a mixture of epoxy resin (number average molecular weight ≤ 700), which is a reaction product of bisphenol-A-(epichlorhydrin) (50 wt.%) and 1,2,3-Propanetriol, glycidyl ethers (50 wt.%). The specification of L285 is as follows: density (25 °C): 1.18–1.23 g/cm^3, viscosity: 600–900 mPas/25 °C epoxy equivalent: 165–170, epoxy number: 0.59–0.65.

There are different hardeners dedicated for the L285 resin. For the experiment, H287 hardener was selected as characterized by a long gelling time, which makes it easier to prepare fiber reinforced composites. H285 hardener is a mono-constituent substance based on 2,2′-dimethyl-4,4′-methylenebis(cyclohexylamine) with possible impurity in the form of 1,4-bis(butylamino)anthraquinone. The H 287 properties are as follows: density (25 °C): 0.93–0.96 g/cm^3, viscosity: 80–100 mPas/25 °C amine number: 450–500.

For the nanofiller, industrial grade multi-walled nanotubes of 90 wt.%, 10 nm OD, manufactured by Bucky USA (Houston, TX, USA), were chosen. The MWCNTs were delivered in the form of powder. Their properties stated by the manufacturer in the specification were as follows: purity: 90 wt.%, outer diameter (OD): 10–30 nm, inner diameter (ID): 5–10 nm, length: 10–30 μm, specific surface area (SSA): >200 m^2/g, bulk density: 0.06 g/cm^3, true density: ~2.1 g/cm^3.

2.2. Sample Preparation

A pure powder of MWCNTs was used for HR-TEM as well as for reference in Raman spectroscopy. The MWCNTs/epoxy nanocomposite was prepared by direct nanotube incorporation. The mechanical mixing and the ultrasound exposure were implemented. The concentrations of carbon nanotubes were selected so that after the addition of the hardener prepared mass fractions are 0.0, 0.25, 0.5, 0.75, 1.0, and 2.0 wt.%. After that, the composition of L285 and MWCNTs was mixed with H287 in 100:40 resin to hardener weight ratio as recommended by the manufacturer.

To achieve this goal, mixing with a mechanical laboratory stirrer was carried out at the speed of 300 rpm for 10 min. Then, to remove air bubbles, the mixture was placed in a vacuum chamber. The next step was exposure to working frequency of 40 kHz in the Ultron ultrasonic bath for 10 min. Then the composite was casted in silicone molds (Figure 1b) and between sheets of PCV films.

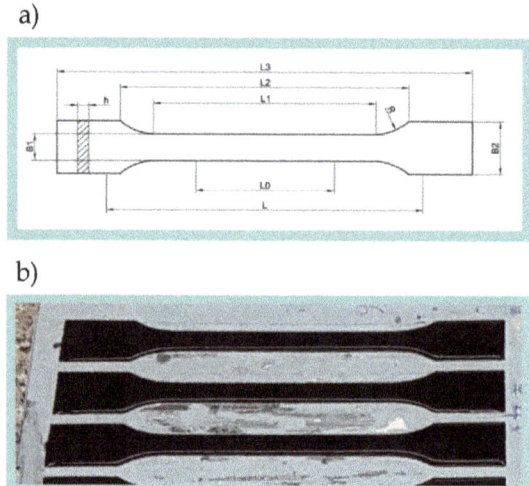

Figure 1. (**a**) Geometry of the sample for tensile test where: L3—Total length: 150 mm; L1—Length of parallel edges narrow zone: 60 mm; R—Radius: 60 mm; B2—Width at ends: 20 mm; B1—Width at narrow zone: 10 mm; h—Thickness: 4 mm; L0—Reference length: 50 mm; L—Length between clamps: 115 mm. (**b**) Nanocomposite casted to the silicone forms.

Specimens for tensile testing, hardness testing, and abrasion testing were cast in the molds. The silicone forms were designed in accordance with the ISO 527-1 standard (Figure 1a). In order to maintain repeatability in the thickness of the cast samples, the upper surface was processed on a numerically controlled milling machine. This procedure also allowed the examination of the core material bypassing the surface layer, which usually has different properties [67]. The composite compressed between sheets of PVC films allowed obtaining flat samples of the material. They were prepared for Raman spectroscopy and for bright field observation in the optical microscope.

3. Methods

3.1. High-Resolution Transmission Electron Microscopy (HR-TEM)

In order to prove that properties of carbon nanotubes are at least as good as those provided by the manufacturer and in order to determine a contaminant and a general composition of the structures, high-resolution transmission electron microscopy (HR-TEM) was performed. HR-TEM was collected on a JEOL (Tokyo, Japan) ARM-200F microscope working at 80 kV equipped with an energy dispersive X-ray (EDX) detector. The samples were prepared by sonicating a solution of MWCNTs powder in pure ethanol. The sonicated product was then drop casted on commercially available Lacey Carbon Cu grids and dried overnight in a vacuum dissector connected to a membrane vacuum pump (<2 mBar).

3.2. Raman Spectroscopy

Raman spectra were obtained using an in Via Ranishaw Raman Microscopy system (Ranishaw, Old Town, Wotton-under-Edge, UK) with a 633 nm He/Ne laser (0.75 mW laser power, Stage I) and 1800 g/mm grating. The laser light was focused on the sample with a 20×/0.75 microscope objective (LEICA, Wetzlar, Germany) All Raman spectra were obtained from 450 to 4000 cm^{-1} using 20 s acquisition time. All spectra were corrected by using the WiRETM 3.3 software attached to the instrument. The measurement of peak positions was performed by using Lorentz profile at OriginPro 8.3 software (Northampton, MA, USA).

3.3. Tensile Testing

The uniaxial tensile testing was carried out on a Zwick Roell (Ulm, Germany) Z5.0 testing machine, operating with the test speed of 2 mm/min and at normal ambient conditions in accordance with ISO 527-1 standard. 10 samples were tested for each mass fraction of MWCNTs. Based on the results, mean values and standard errors were determined.

3.4. Hardness Tests

Shore D hardness was determined according to ISO 868 standard (Bareiss Shore/IRHD Digi Test II, FRT GmbH, Bergisch Gladbach, Germany). 10 samples were tested for each mass fraction of MWCNTs.

3.5. Abrasion Tests

The examination of wear behavior was conducted on TABER Linear Abraser Model 5750 (North Tonawanda, NY, USA) (Figure 2).

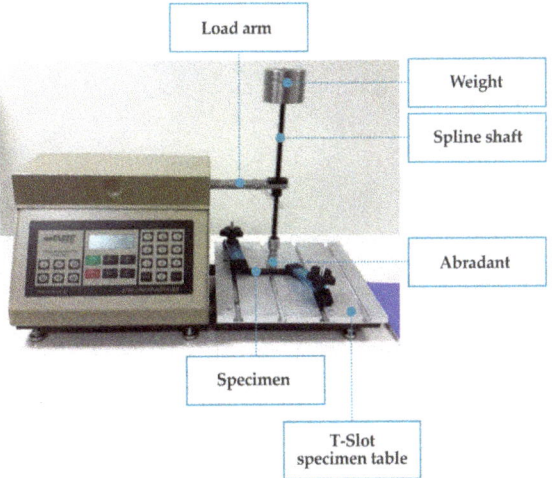

Figure 2. TABER linear abraser model 5750 during the abrasion test.

The conditions during the abrasion test were as follows: stroke length: 50.8 mm, speed: 60 cycles per minute, maximum velocity: 159.52 mm/s, total load: 1850 ± 1 g, abradant model: H-18 (vitrified, medium), abradant diameter: 6.35 mm, maximum number of cycles for one sample: 1000, wear intervals: 200 cycles. Five samples were tested for each mass fraction of MWCNTs.

After each 200-cycle interval, mass loss and topography of the sample were measured. The mass measurements were taken on a precision laboratory weight. The evaluation of the topography was performed on the MicroProf 100 surface metrology tool (FRT GmbH, Bergisch Gladbach, Germany) with CWL 600 μm sensor (FRT GmbH, Bergisch Gladbach, Germany). Measuring characteristics were as follows: measuring range z: 600 μm, resolution (lateral): 2 μm, resolution (vertical): 6 nm. Each time the same part of the specimen surface area was mapped: 9 mm × 4.5 mm rectangle with a long side perpendicular to the abrasion direction. 3D data and profiles were analyzed using FRT Mark III software (FRT GmbH, Bergisch Gladbach, Germany).

3.6. Surface Imaging

Visual inspection was carried out on the Olympus (Tokyo, Japan) BX53M optical microscope with 5× magnification. Bright field method was applied to thin films of composite in order to check whether the obtained image had unfavorable microscale agglomerations or if a gradient arrangement of MWCNTs occurs. The dark field method was applied to casted samples before and during the abrasion test. The application of this method allowed highlighting surface irregularities occurring on the sample. The SEM images were collected on Hitachi (Tokyo, Japan) tabletop microscope model TM3030 plus with accelerating voltage 15 kV and 500× magnification.

4. Results

4.1. HR-TEM

TEM imaging shows MWCNTs outer diameter of 22.3 ± 3.5 nm. The length is variable and difficult to estimate since the tubes are found in bundles. The van der Waals distance between concentrically walls is estimated as 0.289 nm, thus an average of 20 layers is present in the MWCNTs. On the other hand, surface features (marked by yellow arrows), show partial oxidation and damage to the carbon nanotubes. Finally, metallic particles, residual products from the synthesis process, are present and embedded in the carbon nanotubes central sections (Figure 3).

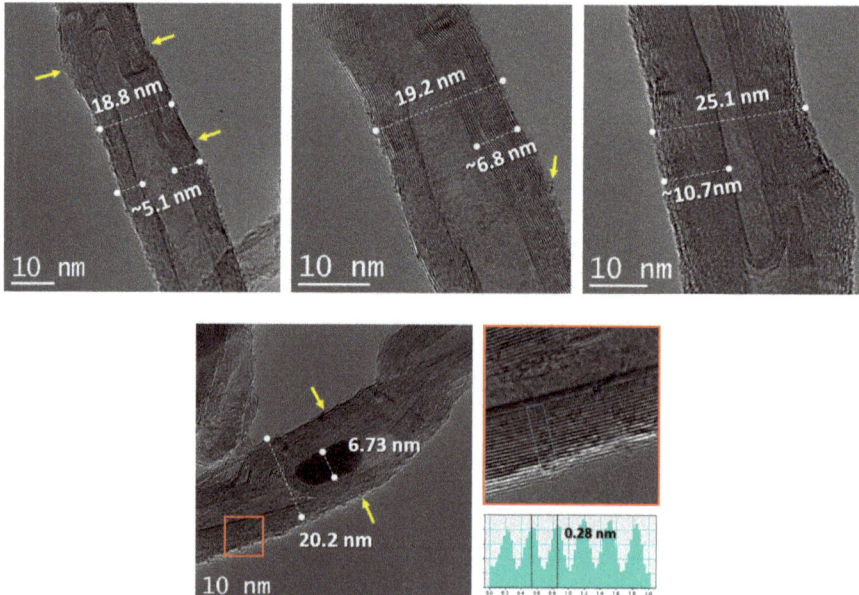

Figure 3. HR-TEM pictures of carbon nanotubes used for MWCNT/epoxy composite.

EDX mapping shows relative oxidation of the MWCNTs surfaces and the FeO_x structure of the central particles. In the oxidized carbon nanotubes, oxygen groups are covalently attached to nanotubes [68]. These are: carboxyl (–COOH), hydroxyl (–OH), epoxy (C–O–C), and carbonyl (C=O).

Traces of nickel and chromium (in the range of 0.5 wt.%) were also detected (Figure 4). The origin of this contamination is connected with the fabrication method of MWCNTs and the values are typical for chemical vapor deposition (CVD) [69], and therefore it was considered that they did not affect the results of other measurements. Large scale collections allowed determining a relative 95 wt.% purity of the MWCNTs with a small concentration of iron and oxygen.

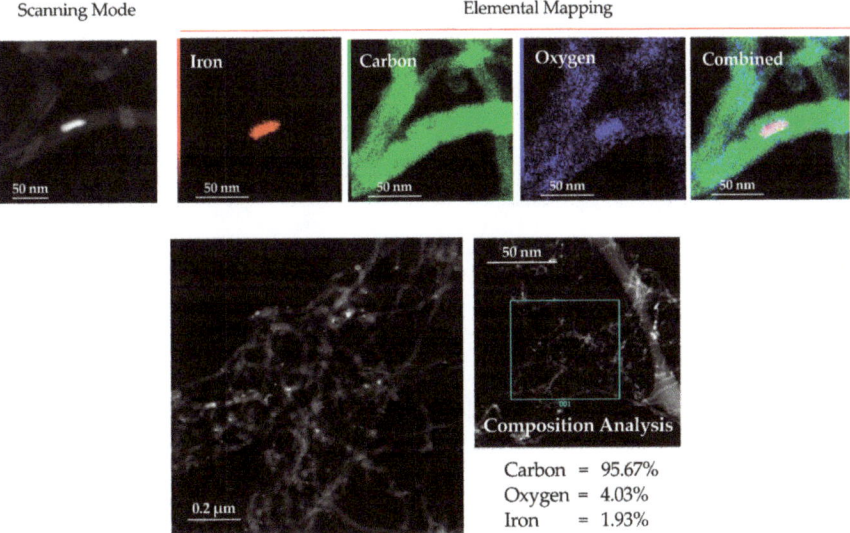

Figure 4. HR-TEM—EDX results of carbon nanotubes used for MWCNT/epoxy composite.

4.2. Raman Spectroscopy

The Raman spectra of 0.25, 0.5, 0.75, 1.0, and 2.0 wt.% MWCNTs content in epoxy matrix and pure reference MWCNTs are presented in Figure 5.

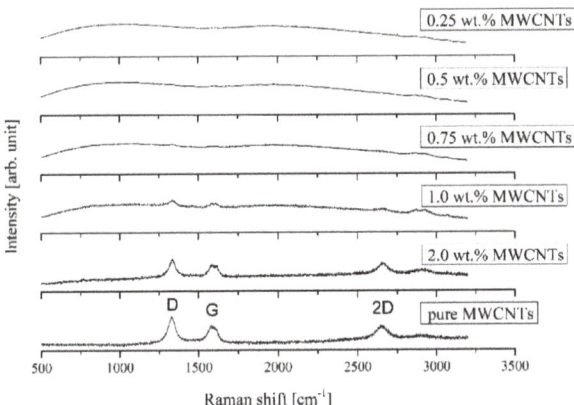

Figure 5. Raman shift for MWCNTs/epoxy composite samples with 0.25, 0.5, 0.75, 1.0, and 2.0 wt.% mass fractions of carbon nanotubes and pure MWCNTs for reference.

The characteristic modes: D (double-resonance mode), G (tangential stretching mode), and 2D (two phonon process) at ~1338 cm^{-1}, ~1586 cm^{-1}, and ~2665 cm^{-1} respectively, confirms the presence of carbon nanotubes in the samples. The D-band is indicative of structural disorder due to disruption of sp^2 C–C bonds, whereas the G-band results come from the tangential vibration of graphitic carbon atoms. All characteristic modes were observed from 0.75 wt.% MWCNTs mass fraction, for lower MWCNTs mass percentage the maxima are visible, although in a fuzzy form, signifying a weaker Raman reflection. Shifting of MWCNTs peak positions was not observed.

4.3. Tensile Testing

Stress–strain curves for all investigated carbon nanotube mass fractions are presented in Figure 6.

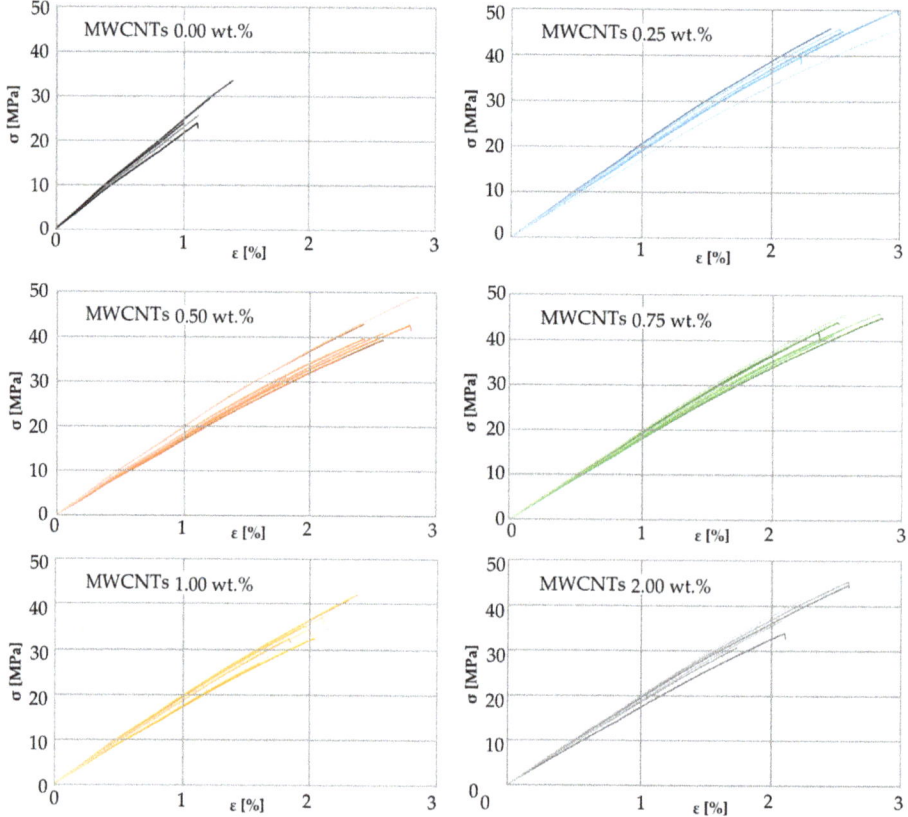

Figure 6. Stress–strain curves for different MWCNTs mass fractions.

For each concentration of MWCNTs, namely 0.0, 0.25, 0.5, 0.75, 1.0, and 2.0 wt.%, a separate graph is presented showing stress–strain curves for all tested samples. The shapes of the stress–strain curves, indicate the brittle fracture of the composites at break, and in this respect all the presented characteristics are similar. As it may be seen form the graphs, a small addition of carbon nanotubes of only 0.25 wt.% causes almost a two-fold increase in tensile strength and a similar increase in relative deformation.

The average ultimate tensile strength σ_u increases for the MWCNTs mass fraction 0.25 wt.% compared to the neat resin by 69% reaching the value of 45.7 MPa and then decreases for higher mass percentages (Figure 7a). Initially increasing and then falling σ_u curve in the concentration range from 0.0 to 0.5 wt.% was also observed in [61] for the polyester matrix samples. Initial growth may be due to arresting and delaying the crack growth effect of MWCNTs, as suggested in [63]. There is a 4th degree polynomial approximation for the eye guiding effect. The interpolated value for 0.1 wt.% is 36% higher than the value for the pure resin, which is only 4% greater than the increase given in [42] but at the same time twice the increase given in [60] for pristine CNTs. It can be noticed that the curve has another extreme a between 1.0 and 2.0 wt.% and begins to grow again, although the whiskers presenting a standard deviation suggest that this rise is of negligible importance.

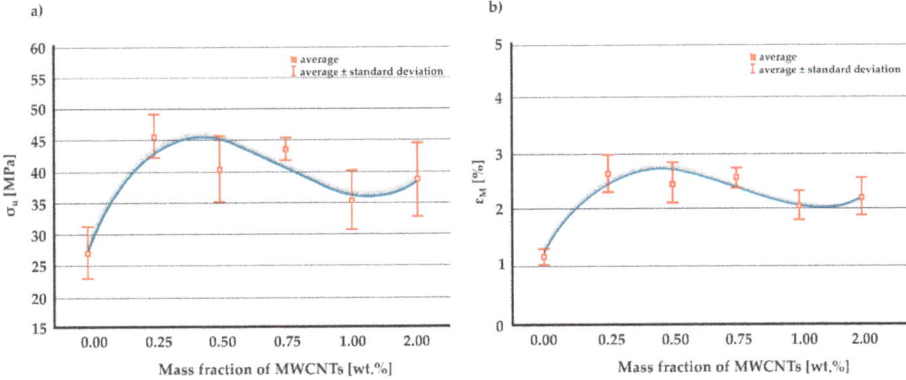

Figure 7. (**a**) Ultimate tensile strength and (**b**) nominal strain at break of the MWCNTs/epoxy composite for different MWCNTs mass fractions.

The curve approximating the change of the nominal strain at break ε_m as a function of carbon nanotubes mass fraction has the same shape as σ_u (Figure 7b). The maximum average value of 2.65% was calculated for MWCNTs 0.25 wt.% which is a 126.5% increase over the resin without a filler. Initially increasing and then falling ε_m curve in the mass fraction range from 0.0 to 0.5 wt.% was also observed in [42].

Calculated mean values of Young's modulus E_t with corresponding standard errors are as follows: 2209 ± 56 MPa for 0.0 wt.%, 1876 ± 27 MPa for 0.25 wt.%, 1728 ± 33 MPa for 0.5 wt.%, 1824 ± 24 MPa for 0.75 wt.%, 1760 ± 31 MPa for 1.0 wt.%, and 1804 ± 38 MPa for 2.0 wt.% MWCNTs. All obtained values are higher than results given in [60], in particular E_t of pure resin is 74% higher, probably due to the use of different Bisphenol-A epoxy resin, namely ED-22. The obtained results mean a significant decrease for a concentration of 0.0 to 0.5 wt.%, an increase for the range of 0.5 to 0.75 wt.% and insignificant changes for MWCNTs mass fraction above 0.75 wt.%. E_t decrease in the concentration range from 0.0 to 0.5 wt.% was also observed in [61] for the polyester matrix samples. The effect of the multi-walled carbon nanotubes concentration on the modulus of elasticity is opposite than on the ultimate tensile strength and the nominal strain at break.

An initial improvement in both mechanical parameters, σ_u and ε_m, is probably due to the effect of an interface between the matrix and the nanofiller. It is also possible that carbon nanotubes transfer the tensile loads through the specimen. A slight deterioration of those properties with the MWCNTs fraction higher than 0.5 wt.% may be a result of a greater concentration of the filler agglomerates. It can be assumed that the change of the toughness is plotted along the similar curve, while the brittleness B (as defined by Brotsow et al. [70]) behaves like inversed toughness (although it is not a rule [71]). The observed increase in toughness may indicate individual nanotubes sliding within the bundles (as noted in [62]). The initial decrease in B (MWCNTs 0.0 wt.% to 0.25 wt.%) could be explained by the improvement of mechanical properties owing to the presence of the filler.

The increase in B (MWCNTs 0.5 wt.% to 1.0 wt.%) could be explained by the impact of the nanofiller concentration on the crosslinking of the polymer. The epoxy resin that is not crosslinked in the whole volume has no tendency to crack. It is probable that the presence of carbon nanotubes catalyzes the crosslinking of the polymer. Another decrease in B may result from increased flexibility of the material due to greater mass fraction of MWCNTs.

4.4. Hardness Test

Shore hardness measurement results have been presented in Figure 8. For a mass fraction of MWCNTs 0.25 wt.%, an average hardness value of 76.2 was obtained. This is a decrease of 7% compared to the value for neat resin. As the mass fraction increases, the hardness increases to values above 83, which is higher than the value for the neat resin, although for 2.0 wt.% it decreases again. It is possible that increasing the hardness in the MWCNTs content ranging from 0.25 wt.% to 0.75 wt.% has the same background as σ_u, as it is discussed on the example of the epoxy-polyamide blend in [72].

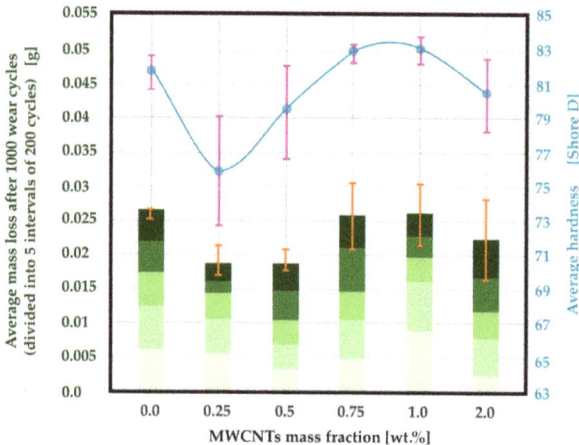

Figure 8. Correlation of mass loss and hardness in relation to MWCNTs mass fraction.

4.5. Abrasion Tests

Figure 9 shows the wear behavior of a MWCNTs 0.25 wt.% sample presented as an example. The diagram shows the mass loss as a function of the number of wear cycles. Each value on the graph is the difference in sample weight before the first 200-cycle interval and the weight after each interval. The slope of the graph therefore corresponds to the value of a weight loss per cycle. The course of the graph is rather linear, which means that the weight losses during subsequent intervals are similar.

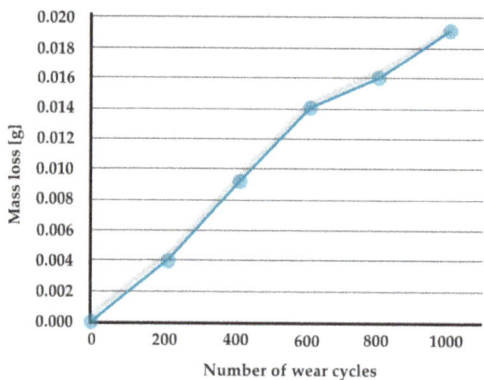

Figure 9. Wear process presented on the example of a MWCNTs 0.25 wt.% sample.

Along with successive intervals of 200 abrasion cycles, the material is gradually penetrated, which is visible on the profiles (Figure 10). As it may be seen on the profiles, volume losses per cycle interval are rather constant and proportional to weight loss

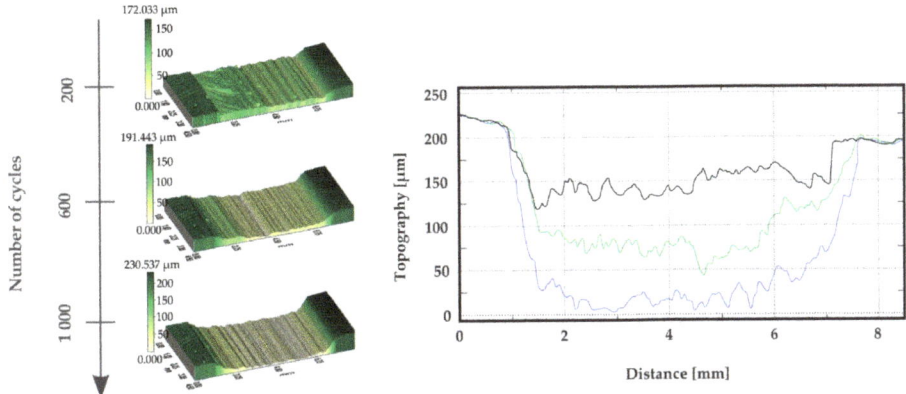

Figure 10. Topography and profiles of MWCNTs 0.25 wt.% sample after 200, 600, and 1000 wear cycles.

The diagram presenting the average mass loss after 1000 wear cycles, with average values marked after intervals of 200 cycles, is very similar to Shore D hardness results (Figure 8). There is a significant drop to 0.019 g for the mass fractions of MWCNTs 0.25 and 0.5 wt.%. This represents a decrease by 28.8% compared to the neat resin. The obtained result contradicts the result presented in [42] where 38% increase in the weight loss was noted. However, the cited article does not contain a detailed description of the research method, and because of that, there is no possibility to determine the cause of the discrepancy. For MWCNTs 0.75 wt.% an increase to the same value as obtained for 0.0 wt.% was noted and, as in the case of the hardness, a decrease occurred for 2.0 wt.%. The wear rate results correspond well with a general decrease in the range from 0.0 wt.% to 2.0 wt.% presented in [10], however, a greater number of investigated MWCNTs mass fraction values revealed the local minimum between 0.25 wt.% and 0.5 wt.%. The superior value of the specific wear rate at 0.5 wt.% exhibits a good agreement with O. Jacobs et al. [28], as well as with [72]. The correlation between the wear rate and the hardness was also observed by Aruniit et al. for the polymer filled with aluminum hydroxide [41].

The increase of friction with hardness in sliding machine element materials was discussed in [73] suggesting the importance of the adhesive component. It is also possible that the harder (and also more brittle) the material is, the more effectively it acts as an abrasive element between the material and the actual abradant after being removed from the nano-composite.

The general improvement in the wear resistance along with nanofiller content may be explained by the MWCNTs playing the role of spacers preventing a close contact between the abrader and the composite surface [74]. In such a case, the MWCNTs would be deformed into a graphene-like lamella, which would decrease the wear loss and the friction coefficient, as shown in [29]. On the contrary, the increase of the average mass loss (for MWCNTs 0.5 wt.% to 0.75 wt.%) could be a result of increased hardness and brittleness B.

Thermal conductivity could also be taken into consideration. As it is known, the influence of the filler on thermal conductivity is very significant [10]. It is possible that a higher thermal conductivity value results in a lower temperature in a sliding contact and therefore improved wear resistance for MWCNTs 2.0 wt.%.

4.6. Surface Imaging

Bright field optical microscopy showed an uneven distribution of MWCNTs and agglomerations in a microscale (Figure 11). As a consequence of the uneven distribution for smaller mass concentrations of nanofiller, bright areas are visible where light shines from under the thin sample.

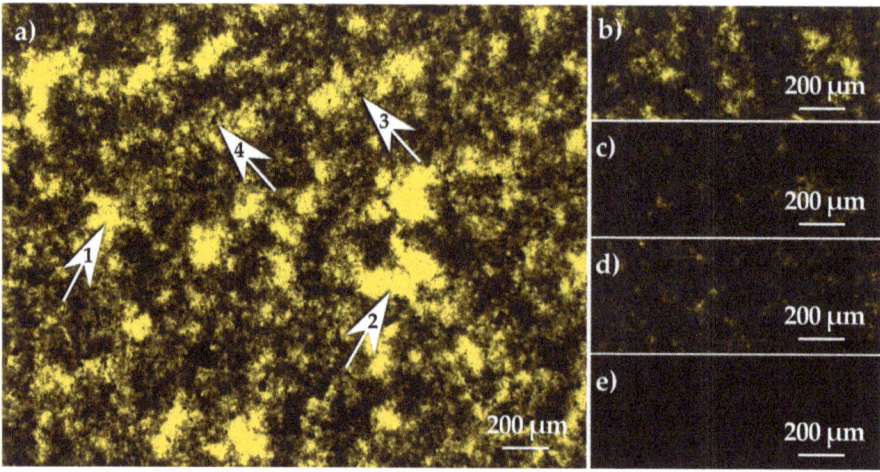

Figure 11. Bright field optical microscopy of MWCNTs (**a**) 0.25, (**b**) 0.5, (**c**) 0.75, (**d**) 0.1, and (**e**) 0.2 wt.%.

The pictures from the dark field microscopy and SEM confirm a large amount of detached material in a form of powder on the abrasive surface of the sample (Figure 12).

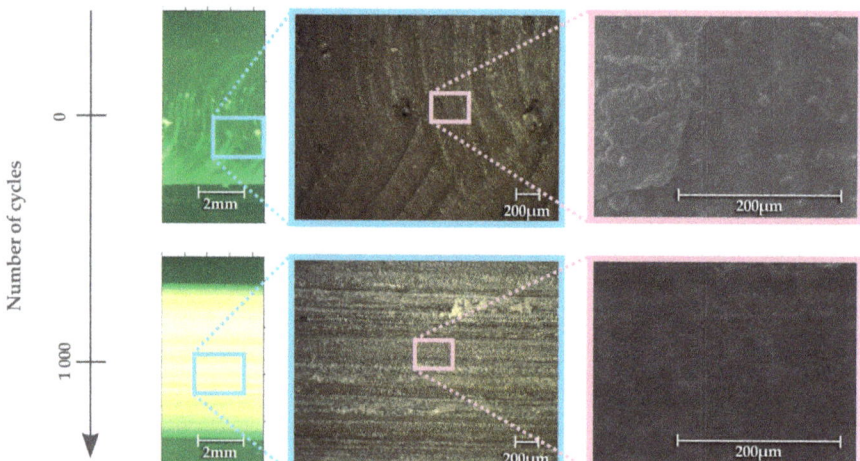

Figure 12. Profilometer topography, optical microscope images and SEM images of MWCNTs 0.25 wt.% sample after 0 and 1000 cycles of abrasive wear.

Before the abrasion test, the characteristic surface obtained due to milling is visible. During the test, the detached material is spread over the surface of the sample, as a result of which the microscopic lines of white powder are visible in the photo from the optical microscope. This powder penetrates surface irregularities. As a result, surface irregularities become less visible in the SEM image compared to the situation before the test. The SEM images show that, according to classification of wear presented in [75], the observed mechanism is mainly adhesive wear.

5. Conclusions

HR-TEM pictures and HR-TEM EDX proved the properties of MWCNTs that occurred to be even better than those declared by the manufacturer. In the manufacturer's specification, the declared purity of coal is 90 wt.%, while the test showed more than 95 wt.%.

Raman spectra, particularly for the highest MWCNT concentration show that carbon nanotubes were not damaged by sonication during the preparation of samples, and that they are present in the nanocomposite samples for all the applied mass fractions.

Optical microscope pictures showed microscopic agglomerations and TEM pictures showed carbon nanotube bundles proving the limitations of dispersivity method based on sonication only.

The wear resistance occurred to be in correlation with ultimate tensile strength. This may mean that the higher the tensile strength, the more difficult it is to tear the material away from the sample causing the mass loss. Moreover, together with tensile strength, maximal strain may influence the wear behavior making the material more deformable in the abrasion area, adapting to the rubbing abradant.

On the other hand, the mass loss and hardness occurred to be in correlation one with the other. For the low mass fraction of MWCNTs (0.25–0.5 wt.%), the mass loss and hardness decreased in comparison to neat resin, while for the higher weight percentage (0.75–1.0 wt.%) an increase in both values was noticed. This may indicate that the lower hardness of the material, the more it tends to form a carbon film covering the contact surfaces and act as a solid lubricant (as presented in [43]).

The study shows that in case of MWCNT-doped epoxy resin within the mass fraction of the nanofiller below 2.0 wt.%, the best wear resistance is achieved for the mass percentage between 0.25 and 0.5. For this reason, such proportions can be promising for structural elements of machines (e.g., aircraft).

Author Contributions: Conceptualization, A.K., M.M. E.K. and T.S.; methodology, A.K., M.S. and T.S.; software, A.K. and M.M.; validation, A.K., M.M., E.K. and M.S.; formal analysis, A.K., M.M. and T.S.; investigation, E.K., M.M., E.C. and M.K.; resources, A.K., M.M. and E.K.; data curation, M.M., E.K., E.C. and M.K.; writing—original draft preparation, M.M., E.C. and M.K.; writing—review and editing, A.K., M.M., T.S., M.S. and E.K.; visualization, E.K. and E.C.; supervision, A.K. and T.S.; project administration, A.K. and M.M.; funding acquisition, A.K. and M.M. All authors have read and agreed to the published version of the manuscript.

Funding: The research was conducted in the framework of the project implemented in 2018–2019, entitled: 'The study of electrical conductivity as a function of fatigue of an aircraft composite material with defined strength properties' No GB/5/2018/209/2018/DA funded by the Ministry of National Defence Republic of Poland. This research was also partly sponsored by the PUT grant 02/25/SBAD/4630.

Acknowledgments: M.M. acknowledges the kind support of Stefan Jurga from CNBM.

Conflicts of Interest: The authors declare no conflicts of interest.

References

1. Pei, X.; Friedrich, K. *Friction and Wear of Polymer Composites*; Elsevier: Amsterdam, The Netherlands, 2016. [CrossRef]
2. Kumar, V.; Sinha, S.K.; Agarwal, A.K. Tribological studies of epoxy composites with solid and liquid fillers. *Tribol. Int.* **2017**, *105*, 27–36. [CrossRef]
3. Bieniaś, J.; Jakubczak, P.; Majerski, K.; Ostapiuk, M.; Surowska, B. Methods of Ultrasonic Testing, as an Effective Way of Estimating Durability and Diagnosing Operational Capability of Composite Laminates Used in Aerospace Industry. *Eksploatacja I Niezawodność* **2013**, *15*, 284–289.
4. Joshi, M.; Chatterjee, U. Polymer nano-composite: An advanced material for aerospace applications. In *Advanced Composite Materials for Aerospace Engineering: Processing, Properties and Applications*; Woodhead Publishing: Cambridge, UK, 2016; pp. 241–264, ISBN 9780081000540.
5. Zhang, X.; Chen, Y.; Hu, J. Recent advances in the development of aerospace materials. *Prog. Aerosp. Sci.* **2018**, *97*, 22–34. [CrossRef]

6. Borowiec, M.; Bochenski, M.; Gawryluk, J.; Augustyniak, M. Analysis of the Macro Fiber Composite Characteristics for Energy Harvesting Efficiency. In *Dynamical Systems: Theoretical and Experimental Analysis*, 2nd ed.; Awrejcewicz, J., Ed.; Springer: Łódź, Poland, 2015; Volume 182, pp. 27–37.
7. Lv, M.; Zheng, F.; Wang, Q.; Wang, T.; Liang, Z. Friction and Wear Behaviours of Carbon and Aramid Fibers Reinforced Polyimide Composites in Simulated Space Environment. *Tribol. Int.* **2015**, *92*, 246–254. [CrossRef]
8. Zhang, M.Q.; Rong, M.Z.; Yu, S.L.; Wetzel, B.; Friedrich, K. Effect of particle surface treatment on the tribological performance of epoxy based nano-composites. *Wear* **2002**, *253*, 1086–1093. [CrossRef]
9. Suresha, B.; Chandramohan, G.; Renukappa, N.M. Siddaramaiah Mechanical and tribological properties of glass–epoxy composites with and without graphite particulate filler. *J. Appl. Polym. Sci.* **2006**, *103*, 2472–2480. [CrossRef]
10. Hussein, S.I.; Abd-Elnaiem, A.M.; Asafa, T.B.; Jaafar, H.I. Effect of incorporation of conductive fillers on mechanical properties and thermal conductivity of epoxy resin composite. *Appl. Phys. A* **2018**, *124*, 475. [CrossRef]
11. Szczepaniak, R.; Rolecki, K.; Krzyzak, A. The influence of the powder additive upon selected mechanical properties of a composite. *IOP Conf. Series: Mater. Sci. Eng.* **2019**, *634*, 012007. [CrossRef]
12. Sławski, S.; Szymiczek, M.; Domin, J. Influence of the reinforcement on the destruction image of the composites panels after applying impact load. *AIP Conf. Proc* **2019**, 020050. [CrossRef]
13. Dydek, K.; Latko-Durałek, P.; Boczkowska, A.; Sałaciński, M.; Kozera, R. Carbon Fiber Reinforced Polymers modified with thermoplastic nonwovens containing multi-walled carbon nanotubes. *Compos. Sci. Technol.* **2019**, *173*, 110–117. [CrossRef]
14. Bellucci, S.; Balasubramanian, C.; Micciulla, F.; Rinaldi, G. CNT composites for aerospace applications. *J. Exp. Nanosci.* **2007**, *2*, 193–206. [CrossRef]
15. Ayatollahi, M.R.; Isfahani, R.B.; Monfared, R.M. Effects of multi-walled carbon nanotube and nanosilica on tensile properties of woven carbon fabric-reinforced epoxy composites fabricated using VARIM. *J. Compos. Mater.* **2017**, *51*, 4177–4188. [CrossRef]
16. Mysiukiewicz, O.; Gospodarek, B.; Ławniczak, P.; Sterzyński, T. Influence of the conductive network creation on electrical, rheological, and mechanical properties of composites based on LDPE and EVA matrices. *Adv. Polym. Technol.* **2018**, *37*, 3542–3551. [CrossRef]
17. Capanidis, D. Selected aspects of the methodology of tribological investigations of polymer materials. *Arch. Civ. Mech. Eng.* **2007**, *7*, 39–55. [CrossRef]
18. Giannakopoulos, A.; Panagiotopoulos, D. Conical indentation of incompressible rubber-like materials. *Int. J. Solids Struct.* **2009**, *46*, 1436–1447. [CrossRef]
19. Krzyzak, A.; Prażmo, J.; Kucharczyk, W. Effect of Natural Ageing on the Physical Properties of Polypropylene Composites. *Adv. Mater. Res.* **2014**, *1001*, 141–148. [CrossRef]
20. Valis, D.; Krzyzak, A. Composite materials reliability assessment and comparison. In *Safety and Reliability of Complex Engineered Systems*; CRC Press: Boca Raton, FL, USA, 2015; pp. 2119–2125.
21. Ruzicka, M.; Dvořák, M.; Schmidová, N.; Šašek, L.; Štěpánek, M. Health and usage monitoring system for the small aircraft composite structure. *AIP Conf. Proc.* **2017**, *1862*, 20007. [CrossRef]
22. Oliwa, R.; Oleksy, M.; Oliwa, J.; Wegier, A.; Krauze, S.; Kowalski, M. Fire resistant glass fabric-epoxy composites with reduced smoke emission. *Polimery* **2019**, *64*, 290–293. [CrossRef]
23. Oliwa, R.; Oleksy, M.; Czech-Polak, J.; Płocińska, M.; Krauze, S.; Kowalski, M. Powder-epoxy resin/glass fabric composites with reduced flammability. *J. Fire Sci.* **2019**, *37*, 155–175. [CrossRef]
24. Policandriotes, T.; Filip, P. Effects of selected nanoadditives on the friction and wear performance of carbon-carbon aircraft brake composites. *Wear* **2011**, *271*, 2280–2289. [CrossRef]
25. Karaeva, A.R.; Kazennov, N.V.; Urvanov, S.A.; Zhukova, E.A.; Mordkovich, V. Carbon Fiber-Reinforced Polyurethane Composites with Modified Carbon–Polymer Interface. *Proc. Sci. Prac. Conf. Res. Dev.* **2017**, 415–420. [CrossRef]
26. Setua, D.K.; Mordina, B.; Srivastava, A.K.; Roy, D.; Prasad, N.E. Carbon nanofibers-reinforced polymer nanocomposites as efficient microwave absorber. In *Fiber-Reinforced Nanocomposites: Fundamentals and Applications*; Elsevier Science: Amsterdam, The Netherlands, 2020; pp. 395–430. [CrossRef]
27. Jia, Z.; Hao, C.; Yan, Z.; Yang, Y. Effects of Nanoscale Expanded Graphite on the Wear and Frictional Behaviours of Polyimide-Based Composites. *Wear* **2015**, *338–339*, 282–287. [CrossRef]

28. Jacobs, O.; Xu, W.; Schadel, B.; Wu, W. Wear behaviour of carbon nanotube reinforced epoxy resin composites. *Tribol. Lett.* **2006**, *23*, 65–75. [CrossRef]
29. Sakka, M.M.; Antar, Z.; Elleuch, K.; Feller, J.F. Tribological response of an epoxy matrix filled with graphite and/or carbon nanotubes. *Friction* **2017**, *5*, 171–182. [CrossRef]
30. Singh, N.P.; Gupta, V.; Singh, A.P. Graphene and carbon nanotube reinforced epoxy nanocomposites: A review. *Polymer* **2019**, *180*, 121724. [CrossRef]
31. Stobinski, L.; Lesiak, B.; Kövér, L.; Toth, J.; Biniak, S.; Trykowski, G.; Judek, J. Multiwall carbon nanotubes purification and oxidation by nitric acid studied by the FTIR and electron spectroscopy methods. *J. Alloy. Compd.* **2010**, *501*, 77–84. [CrossRef]
32. Stobinski, L.; Lesiak, B.; Zemek, J.; Jiricek, P.; Biniak, S.; Trykowski, G. Studies of oxidized carbon nanotubes in temperature range RT–630°C by the infrared and electron spectroscopies. *J. Alloy. Compd.* **2010**, *505*, 379–384. [CrossRef]
33. Stobinski, L.; Lesiak, B.; Zemek, J.; Jiricek, P. Time dependent thermal treatment of oxidized MWCNTs studied by the electron and mass spectroscopy methods. *Appl. Surf. Sci.* **2012**, *258*, 7912–7917. [CrossRef]
34. Tjong, S.C. Structural and mechanical properties of polymer nanocomposites. *Mater. Sci. Eng. R: Rep.* **2006**, *53*, 73–197. [CrossRef]
35. Dai, L.; Sun, J. Mechanical Properties of Carbon Nanotubes-Polymer Composites. In *Carbon Nanotubes—Current Progress of Their Polymer Composites*; InTechOpen: London, UK, 2016; Available online: https://www.intechopen.com/books/carbon-nanotubes-current-progress-of-their-polymer-composites/mechanical-properties-of-carbon-nanotubes-polymer-composites (accessed on 12 May 2020).
36. Kashyap, A.; Singh, N.P.; Arora, S.; Singh, V.; Gupta, V. Effect of amino-functionalization of MWCNTs on the mechanical and thermal properties of MWCNTs/epoxy composites. *Bull. Mater. Sci.* **2020**, *43*, 1–9. [CrossRef]
37. Francisco, W.; Ferreira, F.V.; Cividanes, L.D.S.; Coutinho, A.D.R.; Thim, G. Functionalization of Multi-Walled Carbon Nanotube and Mechanical Property of Epoxy-Based Nanocomposite. *J. Aerosp. Technol. Manag.* **2015**, *7*, 289–293. [CrossRef]
38. Pötschke, P.; Mothes, F.; Krause, B.; Voit, B. Melt-Mixed PP/MWCNT Composites: Influence of CNT Incorporation Strategy and Matrix Viscosity on Filler Dispersion and Electrical Resistivity. *Polymer* **2019**, *11*, 189. [CrossRef] [PubMed]
39. Paun, C.; Obreja, C.; Comanescu, F.; Tucureanu, V.; Tutunaru, O.; Romanitan, C.; Ionescu, O. Epoxy nanocomposites based on MWCNT. In Proceedings of the 2019 International Semiconductor Conference (CAS), Sinaia, Romania, 9–11 October 2019; pp. 237–240.
40. Friedrich, K. Polymer composites for tribological applications. *Adv. Ind. Eng. Polym. Res.* **2018**, *1*, 3–39. [CrossRef]
41. Aruniit, A.; Antonov, M.; Kers, J.; Krumme, A. Determination of Resistance to Wear of Particulate Composite. *Key Eng. Mater.* **2014**, *604*, 188–191. [CrossRef]
42. Kharitonov, A.; Tkachev, A.; Blohin, A.; Dyachkova, T.; Kobzev, D.; Maksimkin, A.; Mostovoy, A.; Alekseiko, L. Reinforcement of Bisphenol-F epoxy resin composites with fluorinated carbon nanotubes. *Compos. Sci. Technol.* **2016**, *134*, 161–167. [CrossRef]
43. Zhang, Z.; Wu, T.; Xie, Y.; Liu, J. Effect of carbon nanotubes on friction and wear of a piston ring and cylinder liner system under dry and lubricated conditions. *Friction* **2016**, *5*, 147–154. [CrossRef]
44. Ren, Z.; Yangyu, Y.; Lin, Y.; Guo, Z. Tribological Properties of Molybdenum Disulfide and Helical Carbon Nanotube Modified Epoxy Resin. *Materials* **2019**, *12*, 903. [CrossRef]
45. Oliwa, R.; Oliwa, J.; Bulanda, K.; Oleksy, M.; Budzik, G. Effect of modified bentonites on the crosslinking process of epoxy resin with alifphatic amine as curing agent. *Polimery* **2019**, *64*, 499–503. [CrossRef]
46. Kozioł, M.; Jesionek, M.; Szperlich, P. Addition of a small amount of multiwalled carbon nanotubes and flaked graphene to epoxy resin. *J. Reinf. Plast. Compos.* **2017**, *36*, 640–654. [CrossRef]
47. Smith, D.J. High Resolution Transmission Electron Microscopy. In *Handbook of Microscopy for Nanotechnology*; Yao, N., Wang, Z.L., Eds.; Springer: Boston, MA, USA, 2015; pp. 427–453.
48. Özden, S.; Narayanan, T.N.; Tiwary, C.S.; Dong, P.; Hart, A.H.C.; Vajtai, R.; Ajayan, P.M. 3D Macroporous Solids from Chemically Cross-linked Carbon Nanotubes. *Small* **2014**, *11*, 688–693. [CrossRef]
49. Image Modes in TEM—Lattice Images. Available online: https://www.microscopy.ethz.ch/TEM_HRTEM.htm (accessed on 10 October 2019).

50. Manrique, J.A.; Marí, B.; Ribes-Greus, A.; Monreal, L.; Teruel, R.; Gascon, M.L.; Sans, J.; Marí-Guaita, J. Study of the Degree of Cure through Thermal Analysis and Raman Spectroscopy in Composite-Forming Processes. *Materials* **2019**, *12*, 3991. [CrossRef] [PubMed]
51. Rycewicz, M.; Macewicz, Ł.; Kratochvil, J.; Stanisławska, A.; Ficek, M.; Sawczak, M.; Stranak, V.; Szkodo, M.; Bogdanowicz, R. Physicochemical and Mechanical Performance of Freestanding Boron-Doped Diamond Nanosheets Coated with C:H:N:O Plasma Polymer. *Materials* **2020**, *13*, 1861. [CrossRef] [PubMed]
52. Krishna, R.; Unsworth, T.; Edge, R. Raman Spectroscopy and Microscopy. In *Reference Module in Materials Science and Materials Engineering*; Elsevier BV: Oxford, UK, 2016.
53. Lawson, E.E.; Barry, B.W.; Williams, A.C.; Edwards, H.G.M. Biomedical Applications of Raman Spectroscopy. *J. Raman Spectrosc.* **1997**, *28*, 111–117. [CrossRef]
54. Kline, N.J.; Treado, P.J. Raman Chemical Imaging of Breast Tissue. *J. Raman Spectrosc.* **1997**, *28*, 119–124. [CrossRef]
55. Smith, G.D.; Clark, R. Raman microscopy in archaeological science. *J. Archaeol. Sci.* **2004**, *31*, 1137–1160. [CrossRef]
56. Nasdala, L.; Smith, D.C.; Kaindl, R.; Ziemann, M.A. *Raman Spectroscopy: Analytical Perspectives in Mineralogical Research*; Eötvös University Press: Budapest, Hungary, 2004; ISBN-10:352728138X. [CrossRef]
57. Edwards, H.G. Spectroscopy, Raman. In *Digital Encyclopedia of Applied Physics*; Wiley: Hoboken, NJ, USA, 2005; ISBN-10:352728138X.
58. Ramana, G.V.; Padya, B.; Kumar, R.N.; Prabhakar, K.V.P.; Jain, P.K. Mechanical properties of multi-walled carbon nanotubes reinforced polymer nanocomposites. *Indian J. Eng. Mater. Sci.* **2010**, *17*, 331–337.
59. Liu, L.; Wagner, H.D. Rubbery and glassy epoxy resins reinforced with carbon nanotubes. *Compos. Sci. Technol.* **2005**, *65*, 1861–1868. [CrossRef]
60. Blokhin, A.N.; Dyachkova, T.P.; Maksimkin, A.V.; Stolyarov, R.A.; Suhorukov, A.K.; Burmistrov, I.N.; Kharitonov, A.P. Polymer composites based on epoxy resin with added carbon nanotubes. *Full Nanotub. Carbon Nanostruct.* **2019**, *28*, 45–49. [CrossRef]
61. Shokrieh, M.; Saeedi, A.; Chitsazzadeh, M. Mechanical properties of multi-walled carbon nanotube/polyester nanocomposites. *J. Nanostruct. Chem.* **2013**, *3*, 20. [CrossRef]
62. Marouf, B.T.; Mai, Y.-W.; Bagheri, R.; Pearson, R.A. Toughening of Epoxy Nanocomposites: Nano and Hybrid Effects. *Polym. Rev.* **2016**, *56*, 70–112. [CrossRef]
63. Hosur, M.; Mahdi, T.H.; E Islam, M.; Jeelani, S. Mechanical and viscoelastic properties of epoxy nanocomposites reinforced with carbon nanotubes, nanoclay, and binary nanoparticles. *J. Reinf. Plast. Compos.* **2017**, *36*, 667–684. [CrossRef]
64. Lin, Y.; Lafarie-Frenot, M.C.; Bai, J.; Gigliotti, M. Numerical simulation of the thermoelectric behaviour of CNTs/CFRP aircraft composite laminates. *Adv. Aircr. Space Sci.* **2018**, *5*, 633–652.
65. Shen, S.; Yang, L.; Wang, C.; Wei, L. Effect of CNT orientation on the mechanical property and fracture mechanism of vertically aligned carbon nanotube/carbon composites. *Ceram. Int.* **2020**, *46*, 4933–4938. [CrossRef]
66. Krzyzak, A.; Mucha, M.; Pindych, D.; Racinowski, D. Analysis of abrasive wear of selected Aircraft materials in various abrasion conditions. *J. KONES Powertrain Transp.* **2018**, *25*, 217–222.
67. Musiał, J.; Horiashchenko, S.; Polasik, R.; Musiał, J.; Kałaczyński, T.; Matuszewski, M.; Śrutek, M. Abrasion Wear Resistance of Polymer Constructional Materials for Rapid Prototyping and Tool-Making Industry. *Polymer* **2020**, *12*, 873. [CrossRef]
68. Maciejewska, B.M.; Jasiurkowska-Delaporte, M.; Vasylenko, A.I.; Koziol, K.K.; Jurga, S. Experimental and theoretical studies on the mechanism for chemical oxidation of multiwalled carbon nanotubes. *RSC Adv.* **2014**, *4*, 28826–28831. [CrossRef]
69. Kumar, M.; Ando, Y. Chemical vapor deposition of carbon nanotubes: A review on growth mechanism and mass production. *J. Nanosci. Nanotechnol.* **2010**, *10*, 3739–3758. [CrossRef]
70. Brostow, W.; Lobland, H.E.H.; Narkis, M. Sliding wear, viscoelasticity, and brittleness of polymers. *J. Mater. Res.* **2006**, *21*, 2422–2428. [CrossRef]
71. Brostow, W.; Lobland, H.E.H.; Narkis, M. The concept of materials brittleness and its applications. *Polym. Bull.* **2011**, *67*, 1697–1707. [CrossRef]
72. Al Shaabania, Y.A. Wear and Friction Properties of Epoxy- Polyamide Blend Nanocomposites Reinforced by MWCNTs. *Energy Procedia* **2019**, *157*, 1561–1567. [CrossRef]

73. Kalacska, G. An engineering approach to dry friction behaviour of numerous engineering plastics with respect to the mechanical properties. *Express Polym. Lett.* **2013**, *7*, 199–210. [CrossRef]
74. Krzyzak, A.; Kosicka, E.; Szczepaniak, R.; Szymczak, T. Evaluation of the properties of polymer composites with carbon nanotubes in the aspect of their abrasive wear. *J. Achiev. Mat. Manu. Eng.* **2019**, *1*, 5–12. [CrossRef]
75. Kato, K. Classification of wear mechanisms/models. *Proc. Inst. Mech. Eng. Part J J. Eng. Tribol.* **2002**, *216*, 349–355. [CrossRef]

© 2020 by the authors. Licensee MDPI, Basel, Switzerland. This article is an open access article distributed under the terms and conditions of the Creative Commons Attribution (CC BY) license (http://creativecommons.org/licenses/by/4.0/).

Article

Influence of Carbon Nanotubes on the Mechanical Behavior and Porosity of Cement Pastes Prepared by A Dispersion on Cement Particles in Isopropanol Suspension

Vanessa Vilela Rocha * and Péter Ludvig

Department of Civil Engineering, Federal Center for Technological Education of Minas Gerais, Belo Horizonte, Minas Gerais 30.421-169, Brazil; peter@cefetmg.br
* Correspondence: vanessa.vilela@cefetmg.br

Received: 2 April 2020; Accepted: 15 June 2020; Published: 15 July 2020

Abstract: Cement composites prepared with nanoparticles have been widely studied in order to achieve superior performance structures. The incorporation of carbon nanotubes (CNTs) is an excellent alternative due to their mechanical, electrical, and thermal properties. However, effective dispersion is essential to ensure strength gains. In the present work, cement pastes were prepared incorporating CNTs in proportions up to 0.10% by weight of cement, dispersed on the surface of anhydrous cement particles in isopropanol suspension and using ultrasonic agitation. Digital image correlation was employed to obtain basic mechanical parameters of three-point bending tests. The results indicated a 34% gain in compressive strength and 12% in flexural tensile strength gains, respectively, as well as a 70% gain in fracture energy and 14% in fracture toughness in the presence of 0.05% CNTs were recorded. These results suggest that CNTs act as crack propagation controllers. Moreover, CNT presence contributes to pore volume reduction, increases the density of cement pastes, and suggests that CNTs additionally act as nucleation sites of the cement hydration products. Scanning electron microscopy images indicate effective dispersion as a result of the methodology adopted, plus strong bonding between CNTs and the cement hydration product. Therefore, CNTs can be used to obtain more resistant and durable cement-based composites.

Keywords: carbon nanotubes; dispersion; cement

1. Introduction

Carbon nanotubes (CNTs) have been attracting the attention of the scientific community over the last few decades due to their extraordinary mechanical, thermal, and electrical properties. The CNTs' mechanical properties, such as high tensile strength and Young's modulus, are the principal reasons for their use as composite materials [1]. In order to achieve effective composite preparation, good dispersion and homogeneous distribution of CNTs are essential. The hydrophobic behavior causes CNT agglomerations and the clusters formed by their presence in high amounts can compromise the mechanical properties [2–4]. In addition, due to this characteristic, the nanostructured cement composites' strength presents high variability [2], and consequently, reproducibility and statistical analysis are important to the validation of results.

Studies on CNTs in cement-based materials have suggested improvements in mechanical properties of cement pastes [5–11], including reduction of porosity [4,710], reduction of early shrinkage [11], an increase in the fracture energy [11,12], and an increase in fracture toughness and proposal that the presence of well-dispersed CNTs in the cement matrix makes it necessary to apply higher stress to cause cracking and specimen rupture [6]. Moreover, CNTs contributed to a refinement of the mesopores and resulted in a denser matrix [13,14]. Cement paste, with up to 0.20% of CNTs dispersed in aqueous surfactant solution by sonication, indicated that the quantity of macropores (diameter greater than 50 nm) was reduced, while the quantity of pores smaller than 50 nm was increased [15]. Macropores in hardened cement composites are considered to be detrimental as they cause worse mechanical properties and lower density of these materials [15]. Mortars prepared with CNTs dispersed in an aqueous solution with superplasticizers registered a reduction in the quantity of micropores (less than 2 nm) and macropores (greater than 50 nm). It is alleged that this pore refinement on a micro and macro scale is more relevant to obtain gains in mechanical strength than reducing the total porosity [16].

With the aim to increase the dispersion efficiency of CNTs, nanomaterial can be submitted to chemical treatments, which is called functionalization. Functionalization occurs through noncovalent (weak bonds with CNTs) and covalent (strong interactions, usually with significant modifications on CNT properties) interactions [17]. Functionalization treatments result in good dispersion of CNTs in aqueous solutions [18], however, researchers suggest that some types of functionalization can damage the carbon nanotubes structure [19,20].

According to Alsharefa et al. (2017) [21], dispersion through chemical mixtures damaged the structure of CNTs and studies involving physical dispersion should be intensified. Thereby, an option of a non-functionalized methodology through non-aqueous suspensions has been adopted for effective dispersion of CNTs in cement particles [1,22]. Isopropanol, which is a less polar solvent as compared with water, is capable of promoting superior CNT dispersion on the surface of cement particles. However, it does not cause any hydration reaction in Portland cement. The dispersion is enhanced by the employment of ultrasonic frequency agitation and mechanical shaking. Isopropanol is removed thereafter, followed by the preparation of the cement paste. This process can be considered much simpler in execution than the use of covalent functionalization. At the same time, the advantages of this methodology are evidenced by the densification of hydrated calcium silicates (C–S–H) [22], which have the major responsibility for the final strength of cement-based materials. Furthermore, this dispersion process provides an indication of a strong connection between CNTs and hydrated cement matrix, controlling the crack propagation [23]. The dispersion of similar nanomaterial, i.e., multilayer graphene, in isopropanol results in adequate dispersion, avoiding multilayer graphene agglomeration and recording improvements in mechanical properties of the composites [24]. The nanocomposite produced with multilayer graphene shows approximately 100% higher splitting tensile and compressive strengths.

This present article aims to evaluate the effect of adding CNTs on the mechanical properties of cement. In order to achieve the objectives, predispersions in isopropanol media of CNTs on cement particles were prepared and used to evaluate the physical and mechanical behavior of cement pastes incorporating 0%, 0.05%, and 0.10% of CNTs according to cement weight. Using these proportions and methodology, Rocha and Ludvig (2018) [9] achieved approximately 50% gain in compressive and splitting tensile strengths at 0.05% of CNTs, suggesting that the optimum range for incorporation of CNTs based on that methodology is close to this ratio. Hawreen et al. (2018) [3] also confirmed that amounts of CNTs, up to 0.10%, were more effective for increasing the flexural and compressive strengths.

2. Materials and Methods

The cement pastes were prepared without additives or admixtures, and the CNTs did not undergo any type of functionalization. At 28 days, the cement pastes were mechanically characterized regarding their compressive strength, flexural tensile strength, fracture energy, and fracture toughness. Moreover, their void index and pore distribution were determined, as well as their thermogravimetric analysis was performed.

2.1. Materials

The materials used in this research were:

- Multiwalled carbon nanotube (MWCNT) with estimated tube lengths between 5 and 30 μm, 99% of external diameter between 10 and 50 nm, and purity higher than 93%, produced by CTNano, Belo Horizonte, Minas Gerais, Brazil;
- Brazilian type CP-V Portland cement because of its low percentage of mineral additions;
- Isopropanol absolute grade;
- Potable water for cement hydration.

2.2. Dispersion Process

Two different formulations with CNTs were carried out to prepare the cement paste, i.e., adding 0.05% and 0.10% by total weight of cement (bwoc).

The CNTs were dispersed in non-aqueous isopropanol solution. The steps for dispersing CNTs in anhydrous cement particles are described in Figure 1.

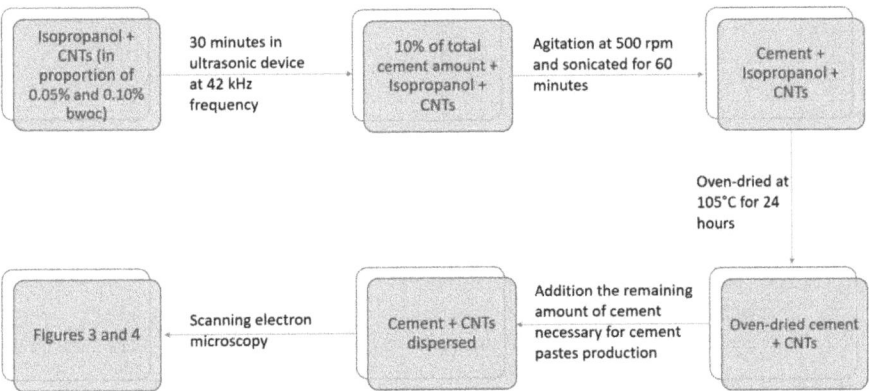

Figure 1. The adopted methodology to disperse carbon nanotubes (CNTs) in a suspension of isopropanol.

The scanning electron microscopy images were acquired using a FEG Quanta 200 made by FEI Company (located at the Center of Microscopy, UFMG, Belo Horizonte, Minas Gerais, Brazil) to evaluate the CNT dispersion. To ensure the electrical conduction on the samples, a 5 nanometers thick carbon coating was applied.

The surface area of the anhydrous cements with dispersed CNTs with concentrations of 0%, 0.05%, and 0.10%, were also determined by nitrogen adsorption using multipoint technique.

2.3. Evaluation of the Mechanical Behavior

The cement pastes were prepared in a mortar blender with 0.33 water/cement ratio. For each formulation, 4 prismatic specimens with 4 cm height, 4 cm width, and 16 cm length for flexural tensile test were prepared. The specimens were maintained in lime-saturated water until they reached 28 days, and then tested. The formulations analyzed are described in Table 1.

Table 1. Cement paste formulations.

Identification	Formulation
REF-ISO	Cement paste prepared without CNTs
0.05-ISO	Cement paste prepared with addition of 0.05% of CNTs
0.10-ISO	Cement paste prepared with addition of 0.10% of CNTs

Prior to the three-point flexural tests, a 1 cm deep notch was made in the middle of the bottom of the beams. The cut was performed in a special apparatus to ensure alignment. In sequence, the lateral sides of the test samples were painted with white and black spray to create a stochastic pattern in each sample to guarantee the digital image correlation (DIC) analysis efficiency. The tests were conducted on a universal mechanical testing machine equipped with 5 kN load cell at a loading speed of 0.25 mm/min.

The flexural test was photographed throughout the loading at 250 milliseconds intervals. Using the sequence of high-resolution and accurate photos, a correlate software recorded the horizontal and vertical displacements suffered during the test, as well as the applied force. The results obtained were used to calculate the fracture energy (G_f) and fracture toughness (K_{IC}) of the samples. These two parameters were calculated according to the formulation indicated by Dally and Riley (1991) [25] and by Hu et al. (2014) [6] as indicated in the equations below:

$$Gf = \frac{m\,g\,\delta + W}{t\,(h-a)} \tag{1}$$

$$K_{IC} = \frac{P\,S}{t\,h^{\frac{3}{2}}} f\left(\frac{a}{h}\right) \tag{2}$$

$$f\left(\frac{a}{h}\right) = 2.9\left(\frac{a}{h}\right)^{\frac{1}{2}} - 4.6\left(\frac{a}{h}\right)^{\frac{3}{2}} + 21.8\left(\frac{a}{h}\right)^{\frac{5}{2}} - 37.6\left(\frac{a}{h}\right)^{\frac{7}{2}} + 38.7\left(\frac{a}{h}\right)^{\frac{9}{2}} \tag{3}$$

where G_f is the fracture energy of the composite; K_{IC} is the fracture toughness of the composite, W is the area under the load-displacement curve, m is the mass of the beam between supports, g is the acceleration due to gravity, δ is the vertical displacement at final failure of the beam, t is the width of the specimen, h is the height of the specimen, a is the depth of the slot, S is the span of the beam, P is the peak load, and f(a/h) is the geometry factor [25].

The results of the flexural tensile strength tests were also compared with the results obtained by Rocha and Ludvig (2018) [9] realized using a splitting tensile strength test.

After the three-point flexural test, one of the remaining halves of the prismatic samples was used to perform compression tests as an adaptation of ASTM C349-02 [26]. The fragments were placed between two metal plates with an area of 4 cm × 4 cm, as indicated in Figure 2. The sample was submitted to continuous loading in the EMIC brand universal equipment located at the CTNano test laboratory, using a 200 kN load cell and load speed of 0.50 mm/sec. The compressive strength results were compared with the results obtained by Rocha and Ludvig (2018) [9].

Figure 2. Setup and specimen failure of the compressive strength test.

It is noteworthy that Rocha and Ludvig (2018) [9] analyzed the mechanical properties based on compressive and splitting tensile tests on 5 cm diameter and 10 cm height cylindrical samples. In the present work, the tensile strength was assessed by flexural test in prismatic samples and the compressive strength was evaluated on the remaining halves of the prismatic samples, as exhibited in Figure 2. Complementing previous work [9], we performed a microstructural analysis of the cement paste to justify the results of mechanical strength, likewise, the mechanical properties analysis was developed according to a different methodology.

2.4. Evaluation of the Physical Parameters

The other half of the prismatic sample was used to determine density in the saturated and dry conditions, void index, and water absorption as follows:

- The samples were dried at 105 °C for 72 h, cooled to room temperature for 30 min, and weighed, obtaining the dry mass (M_S);
- Then, the samples were completely immersed in water and maintained for 72 h;
- After this period, the masses of the submerged saturated sample (M_i) and saturated sample with dry surface (M_{SAT}) were measured.

These indexes were used for the calculations using the follow [27]:

$$Water\ absorption = \frac{M_{SAT} - M_S}{M_S} \times 100 \qquad (4)$$

$$Void\ index = \frac{M_{SAT} - M_S}{M_{SAT} - M_i} \times 100 \qquad (5)$$

$$Density\ of\ dried\ samples = \frac{M_S}{M_{SAT} - M_i} \qquad (6)$$

$$Density\ of\ saturated\ samples = \frac{M_{SAT}}{M_{SAT} - M_i} \qquad (7)$$

Fragments of each paste formulation were subjected to a pore size distribution analysis using nitrogen condensation technique based on DFT (density functional theory), a quantitative assessment of the composition by thermogravimetric analysis, and scanning electron microscopy in order to evaluate the microstructure morphology.

To determine the pore distribution by nitrogen condensation, the cement pastes samples were degassed for 24 h in a vacuum at 30 °C [28]. The analysis was performed using 20 adsorption and 20 desorption points. The pore size distribution was analyzed using the DFT method based on desorption data.

Whereas, to perform the thermogravimetric analysis, the paste fragments were ground and sieved and only the particles smaller than 75 micrometers were subjected to analysis. The analysis was carried out until reaching 1000 °C with a heating rate of 10 °C/min in a synthetic air environment on a Shimadzu Corporation DTA-60H thermal analyzer (located at the CTNano, Belo Horizonte, Minas Gerais, Brazil).

The scanning electron microscopy images were captured using a FEG Quanta 200 FEI equipment (located at the Center of Microscopy, UFMG, Belo Horizonte, Minas Gerais, Brazil) for the hydrated cement pastes with concentrations of 0.05% and 0.10%. Five nanometers carbon coating were used to ensure sample conductivity.

3. Results and Discussions

3.1. Dispersion Process

Figures 3 and 4 show the efficiency of CNT dispersion on cement particles, evaluated by scanning electron microscopy (using a FEG Quanta 200 made by FEI Company equipment located at the Center of Microscopy, UFMG, Belo Horizonte, Minas Gerais, Brazil). The first Figure displays the cement with 0.05% of dispersed CNTs by weight of cement (bwoc), while the second Figure displays the cement with 0.10% CNTs bwoc.

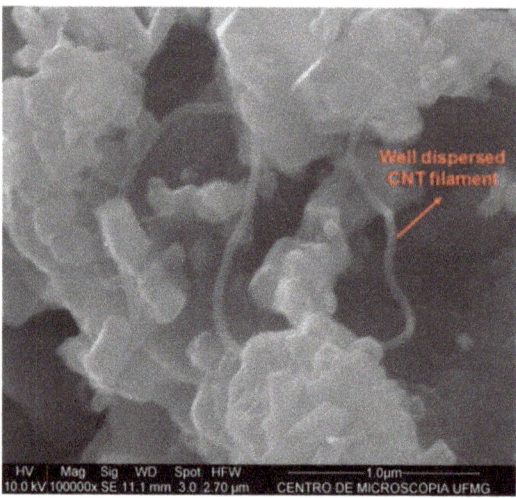

Figure 3. Scanning electron microscopy images of the anhydrous cement with 0.05% of CNTs.

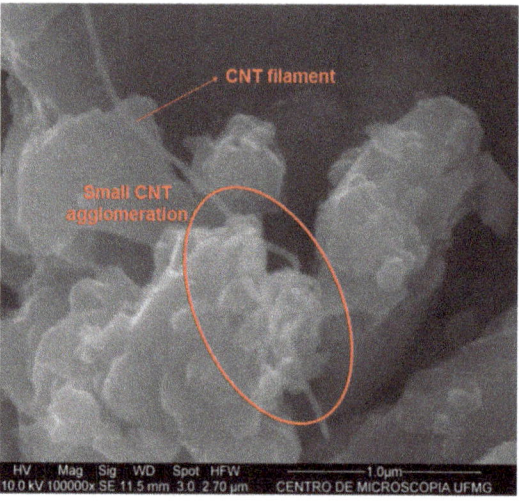

Figure 4. Scanning electron microscopy images of the anhydrous cement with 0.10% of CNTs with small agglomeration.

The CNT filaments are well dispersed in the anhydrous cement with 0.05% concentration of CNTs (see Figure 3). The cement particles with 0.10% of CNTs (Figure 4) also show well-dispersed CNTs on the surface of the cement particles, but small agglomerations are present. Figures 3 and 4 show that the effective CNT dispersion limit for this specific methodology is close to 0.05%, since adding 0.10% of CNTs, small agglomerations are already observed.

The results of the surface area by multipoint nitrogen adsorption technique and the ratio of CNTs dispersed on the anhydrous cement surface area are presented in Table 2, and as expected, the presence of CNTs increased the surface area due to their high specific surface area. Incorporating 0.05% of CNTs, the surface area increased 4.4%, whereas the incorporation of 0.10% recorded an increase of 17.7% as compared with the reference (REF-ISO). The CNT/cement surface area rate was recorded to identify the ideal proportion for CNT filaments to disperse between anhydrous cement particles. According to Figures 3 and 4, CNTs in both proportions of 0.89 and 1.78, respectively, seem to be well dispersed, however, 0.10% of CNTs bwoc (1.78 g/m^2/g) caused some agglomeration.

Table 2. Surface area obtained by nitrogen adsorption.

Identification	Surface Area (m^2 g^{-1})	CNTs / Cement Surface Area Rate (g m^{-2}/g)
Anhydrous Cement	1.687	-
Cement + 0.05% CNT	1.761	0.89
Cement + 0.10% CNT	1.986	1.78

According to the results of Rocha and Ludvig (2018) [9], considering the compressive and tensile strengths, the best result achieved by the dispersion of CNTs in isopropanol, was close to 0.05% bwoc. This implied that the optimal CNT ratio dispersible by that methodology was close to 0.89 g of CNTs per m$^2 \cdot$g^{-1} of anhydrous cement surface area and a superior mechanical performance in cement pastes prepared with anhydrous cement with 0.05% of CNTs was expected.

3.2. Evaluation of the Mechanical Behavior

The results of compressive strength are presented in Figure 5; it is compared with the results achieved by Rocha and Ludvig (2018) [9]. These tests were performed on 5 cm diameter and 10 cm height cylindrical samples (columns on the left side) to determine compressive and splitting tensile strengths, meanwhile, the columns on the right side demonstrate the results of compressive strength determined by the half-prism and flexural tensile strength tests (as described in Section 2.3).

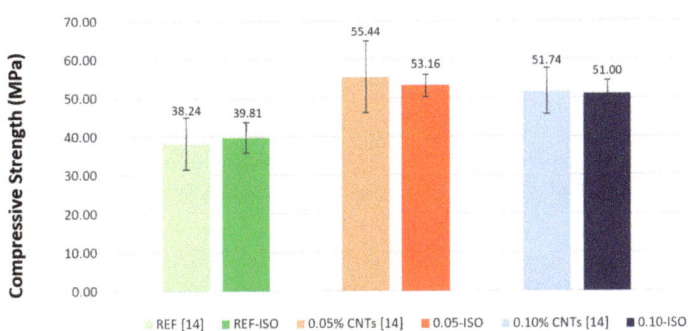

Figure 5. Compressive strength test results as compared with previous results.

The cement paste in the presence of 0.05% and 0.10% of CNTs recorded strength gains of 34% and 28%, respectively, values close to those obtained by Rocha and Ludvig (2018) [9] who recorded 45% and 35%. On the one hand, according to the statistical analysis of variance (ANOVA), the difference between the compressive strength averages of REF, 0.05-ISO and 0.10-ISO pastes can be considered as significant. These gains suggest a reinforcement effect of cement pastes in these CNT proportions. On the other hand, also by ANOVA, the difference between the results for the compressive strength averages in the present work and the results presented by Rocha and Ludvig (2018) [9] are not identified as significant.

The flexural tensile strength results are presented in Figure 6. The compressive strength results are similar to those obtained by Rocha and Ludvig (2018) [9]. In addition to the similar tendencies observed earlier, the flexural tensile strength results are almost two times higher than the splitting tensile strength recorded by the cited authors [9]. The difference can be explained by the fact that flexural tests are more conservative because of the two-dimensional state of stress and the size effect [29–31].

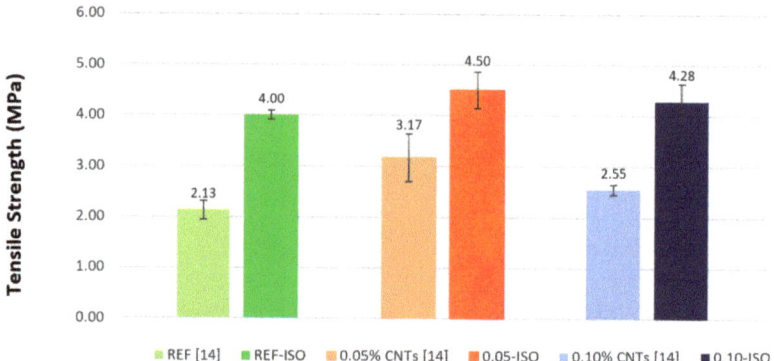

Figure 6. Tensile strength results as compared with previous results.

Cement pastes incorporating 0.05% and 0.10% of CNTs recorded gains of 12% and 7%, respectively, whereas Rocha and Ludvig (2018) [9] recorded 49% and 20%. According to ANOVA, the difference between the averages of tensile by flexural strength is not significant. However, the gains imply a reinforcement of cement pastes as an effect of the addition of predispersed CNTs.

The mechanical characterization by compressive and tensile tests recorded a more expressive increase in cement pastes adding 0.05% of CNTs, which suggests efficient dispersion of the nanomaterial in cement particles by the adopted methodology under this concentration and a possible negative effect at 0.10% addition due to agglomeration.

The results of the fracture energy tests are presented in Figure 7. Cement pastes incorporating 0.05% and 0.10% of CNTs recorded gains of 70% and 35%, respectively. Although the cement paste with 0.05% of CNTs recorded a relatively high standard deviation, the lowest result of fracture energy in the cement paste with carbon nanotubes is still higher than the highest result of the reference paste. According to the ANOVA test, the difference between the fracture energy averages is not significant.

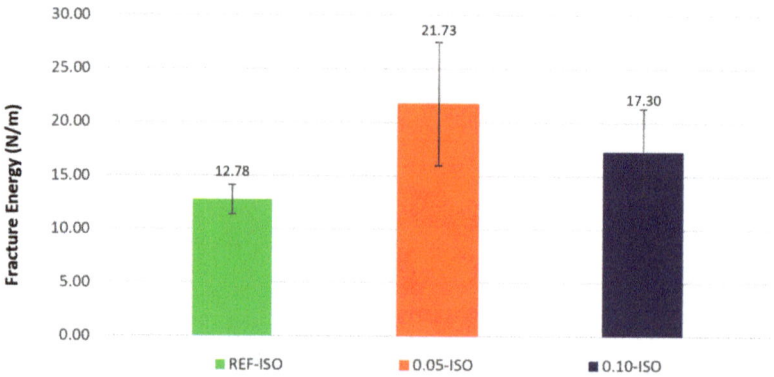

Figure 7. Fracture energy results.

The results of the fracture toughness test are presented in Figure 8. Cement pastes incorporating 0.05% and 0.10% of CNTs recorded gains of 14% and 7%, respectively. According to ANOVA, likewise it is not possible to state that the difference between the averages of fracture toughness is significant, as pointed out in the case of flexural strength and fractural energy. The results of the load-displacement curve integration are presented in Table 3 and the load-displacement diagram of the most representative curve of each cement paste formulation is presented in Figure 9. The results, as presented in the diagram, were extracted from the three-point flexural tests, which were conducted until failure according to the adopted methodology (actuator controlled by vertical displacement and deformations recorded by DIC).

Table 3. Results of the load-displacement curve integration.

Sample	REF-ISO		0.05-ISO		0.10-ISO	
	Area [N.m]	Average (N·m)	Area (N·m)	Average (N.m)	Area (N·m)	Average (N·m)
Specimen 1	0.0174745		0.02981125		0.01956575	
Specimen 2	0.01526125	0.015821167	0.02421901	0.026502003	0.0252675	0.021834167
Specimen 3	0.01472775		0.02547575		0.02066925	

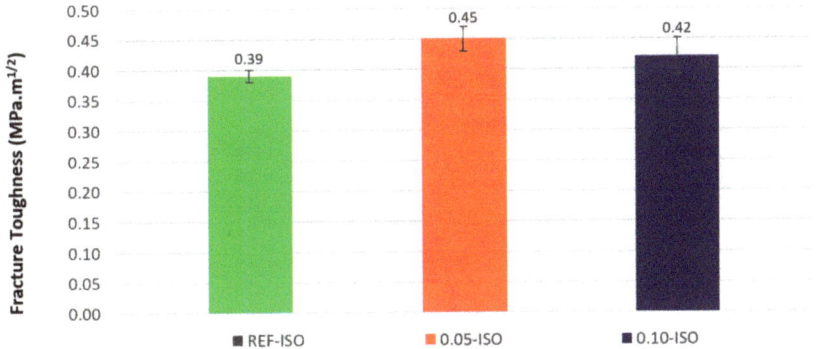

Figure 8. Fracture toughness results.

The fracture energy considering the curve integration area, according to the results obtained by Equation (1), in which a better performance of 0.05-ISO could be observed. This result is expected once the formulation in Equation (1), considering the curve integration area of the load-displacement diagram. The standard deviations of the curve integration areas were 0.0011, 0.0022, and 0.0023 N·m for REF-ISO, 0.05-ISO, and 0.10-ISO, respectively.

The maximum gains in fracture energy and fracture toughness were obtained in the presence of 0.05% of CNTs, which were in accordance with the results of the compression and tensile tests, and suggested that, in such proportions, the hydration products adhered well to the nanomaterial and caused a refinement of the capillary pores [8] and crack control at the submicron level [23]. These results could be related to the effective dispersion of carbon nanotubes that permitted it to act as a fibrous reinforcement and allow greater load application and deformations of the samples before the rupture.

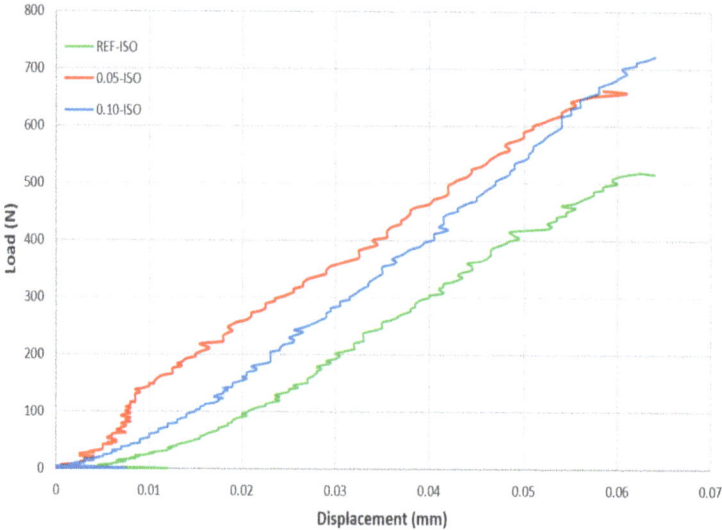

Figure 9. Three-point flexural test load-displacement diagrams.

3.3. Evaluation of the Physical Parameters

The nitrogen adsorption and desorption isotherms and the pore distribution results by DFT analysis are displayed in Figures 10–12, i.e., isotherm, pores diameter results, and pore volume results, respectively.

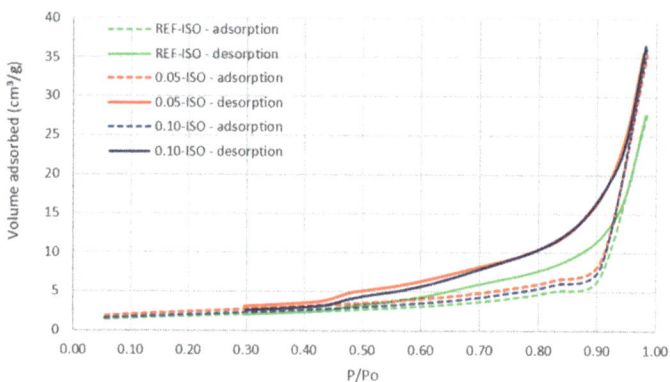

Figure 10. Isotherm curves of cement pastes.

The isotherm curves of these three cement pastes are similar. According to the International Union of Pure and Applied Chemistry (IUPAC), the classification of the three isotherms presented in Figure 10 is isotherm type II and hysteresis type H3, which indicate slit shape pores, characteristics of C–S–H [32].

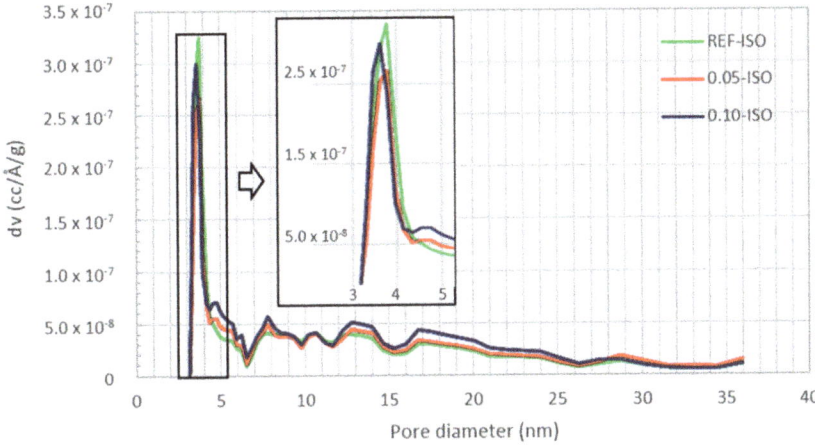

Figure 11. Pore size distribution of cement paste obtained by the density functional theory (DFT) method.

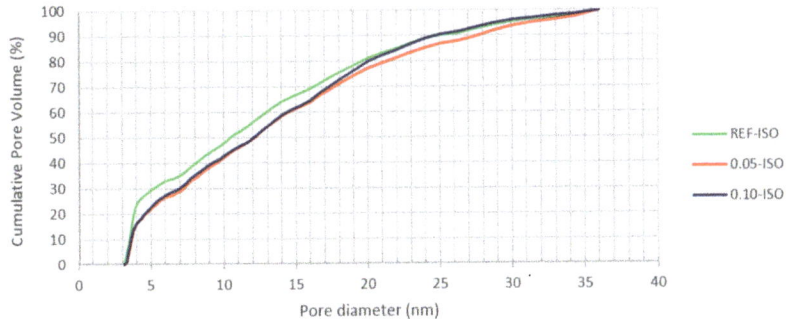

Figure 12. Cumulative pore volume of cement pastes obtained by the DFT method.

According to Figure 10, a higher hysteresis in the presence of CNTs is observed; the desorption curve is more distant from the adsorption for 0.05-ISO and 0.10-ISO, than for the REF-ISO paste. This behavior is related to the difficulty of desorbing the gas molecules that are condensed in the smaller pores identified in cement pastes prepared with CNTs, suggesting pore refinement.

The observed type II isotherm is related to the low degree of pore curvature and the structure [33] corresponding to the characteristics of cement composites.

The pore size distribution graphs obtained by DFT technique were used for the analysis of the pores in the mesoporous region, with diameters between 2 and 50 nm. The results presented in Figures 10 and 11 indicate a higher volume of fine pores in the REF-ISO as compared with cement pastes with CNTs.

According to the results presented in Figure 12, 0.05-ISO and 0.10-ISO have a smaller quantity of pores, up to 20 nm in diameter as compared with REF-ISO. However, the results presented in Figure 10 suggest pore size reduction of 0.05-ISO and 0.10-ISO, which presented difficulty in desorbing the gas. A justification for the desorption difficulty, despite the pore reduction, is the reduction of the specific pores larger than 20 nm in cement pastes with CNTs. These results are in agreement with Xu, Liu and Li (2015) [15], who argued that the presence of CNTs reduced the quantity of pores with diameters larger than 50 nm which were considered to be harmful to the mechanical properties of the cement composites, and increased the quantity of pores smaller than 50 nm. Thereby, the BET results suggest that the presence of CNTs leads to a reduction in pores larger than 20 nm in diameter and an increase in the quantity of pores larger than 20 nm. The reduction of pores that are considered to be harmful contributes to the improvement of the mechanical properties recorded in this work.

The results shown in Figure 12 indicate that 20% of the total pore volume, up to 40 nm (mesoporous range), is higher than the following: (i) 19.2 nm in REF-ISO, (ii) 20.1 nm in 0.05-ISO, and (iii) 21.2 nm in 0.10-ISO. Considering 30% of the total pore volume diameter, the REF-ISO pores are smaller than 5 nm and the 0.05-ISO and 0.10-ISO pores are smaller than approximately 7 nm. These data indicate similarity in the pore distribution of 0.05-ISO and 0.10-ISO pastes and a difference between REF-ISO and pastes prepared with CNTs. The presence of CNTs affected the cement pastes nanostructure resulting in a lower percentage of finer pores, according to Nochaiya and Chaipanich (2010) [14]. As a consequence, cement pastes reinforced with CNTs apparently have a denser matrix, resulting in superior mechanical properties. An explanation for this result could be the CNTs' ability to cause nucleation of the cement hydration products, forming denser structure of the cement paste at the nanoscopic level.

It is noteworthy that the DFT analysis covered pores up to 40 nm and complementary to the nanoscale pore distribution, water absorption, void index, and density of the saturated and dry samples obtained by water absorption technique are presented in Table 4.

Table 4. Water absorption results.

Identification	Water Absorption (%)	Void Index (%)	Density of Saturated Samples (g cm^{-3})	Density of Dried Samples (g cm^{-3})
REF-ISO	20.90%	34.03%	1.97	2.47
0.05-ISO	18.48%	31.96%	2.05	2.54
0.10-ISO	19.23%	32.76%	2.03	2.53

Table 4 demonstrates that the reference presented elevated pore volume obtained by water absorption. Water absorption and void index of the 0.05-ISO and 0.10-ISO pastes exhibit a certain decrease, meanwhile, the density (both saturated and dry) of these samples increase as compared with the reference. These results indicate a lower level of porosity of cement pastes with CNT addition. Furthermore, the water absorption, void index, and density data are consistent with mechanical behavior, as porosity is inversely related to compressive strength. Complementing these results, the pore distribution graphs obtained by the DFT technique analyzed the mesopores and recorded that the REF-ISO presented a higher percentage of finer pores. It can be concluded that CNTs affect the pore structure of the cement paste at both scales, reducing the porosity of the C-S-H and the paste. These findings reinforce the hypothesis that CNTs have a nucleating effect on cement hydration products, contributing to the reduction of small pores in this diameter pores range.

The results of thermogravimetric analysis are exhibited in Table 5 and Figure 13.

Table 5. Mass loss recorded by thermogravimetric analysis.

Mass Loss of Each Cement Paste by Temperature Range							
Temperature Range	Decomposition	REF-ISO		0.05-ISO		0.10-ISO	
		Mass (mg)	Mass Loss (%)	Mass (mg)	Mass Loss (%)	Mass (mg)	Mass Loss (%)
30 °C to 150 °C	Water pore	0.341	33.83	0.380	37.34	0.400	39.17
150 °C to 400 °C	Water pore and C-S-H	0.232	22.95	0.242	23,81	0.234	22.92
400 °C to 600 °C	CH	0.192	19.01	0.220	21.60	0.207	20.24
600 °C to 1000 °C	CaCO$_3$	0.244	24.22	0.175	17.25	0.180	17.67

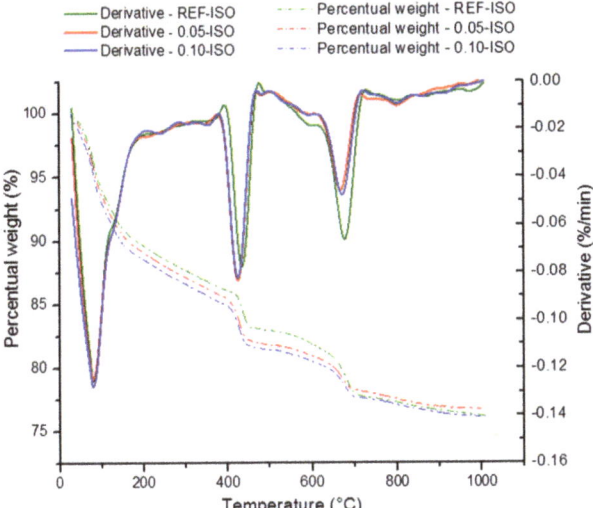

Figure 13. Results of the thermogravimetric analysis.

The ranges indicated in Table 5 are based on the mass loss peaks observed in the derivative thermogravimetric curves. The weight losses recorded are due to the following: (i) up to 150 °C, dehydration of water pore; (ii) from 150 °C to 400 °C, dehydration of different stages of C–S–H; (iii) from 400 °C to 600 °C, dehydroxylation of CH; and (iv) between 600 °C and 1000 °C, decarbonation of $CaCO_3$ [34–36]. There are different concepts on the temperature range of cement paste hydration product decomposition and it is especially challenging to establish a boundary between the water pore and C–S–H decomposition because of the hydrophilic behavior of C-S-H surface [37], therefore, the separation was based on the onset and offset of the mass loss peaks recorded.

The thermogravimetric analysis resulted in similar curve shapes, indicating the decomposition of the same hydration products in the three analyzed samples. The total weight loss was 20.3% for REF-ISO, 19.9% for 0.05-ISO, and 20.4% for 0.10-ISO.

The above results indicate a higher percentage of C–S–H and CH for 0.05-ISO and 0.10-ISO compositions, respectively, which could be explained by the densification of hydration products due to the nucleating effect of CNTs. The third temperature range (600 °C to 1000 °C) corresponds to the calcium carbonate ($CaCO_3$) decomposition and, in this range, the REF-ISO shows a higher amount of $CaCO_3$ as compared with 0.05-ISO and 0.10-ISO. This compound is the result of the carbonic acid (H_2CO_3) reaction, which is formed in the presence of carbon dioxide (H_2CO_3) and humidity (H_2O) in the cement matrix. This result is in accordance with the water absorption test results. The higher pore volume in the REF-ISO would permit higher permeability of CO_2, contributing to a more pronounced formation of $CaCO_3$. The porosity reduction with the addition of CNTs could have occurred because it acts as nucleation sites for hydration products, resulting in C–S–H densification, as suggested by Makar and Chan (2009) [22] and displayed in Table 5. The thermogravimetric analysis evidenced that CNTs in the proportion of 0.05% and 0.10% affect the microstructure of cement pastes, once 0.05-ISO and 0.10-ISO presented higher relative quantity of hydration products (CH and C–S–H) and smaller quantity of $CaCO_3$.

The scanning electron microscopy images of cement pastes with CNTs are shown in Figure 14. Figure 14a,b shows hydrated cement paste with 0.05% of CNTs bwoc and Figure 14c shows hydrated cement paste with 0.10% of CNTs bwoc.

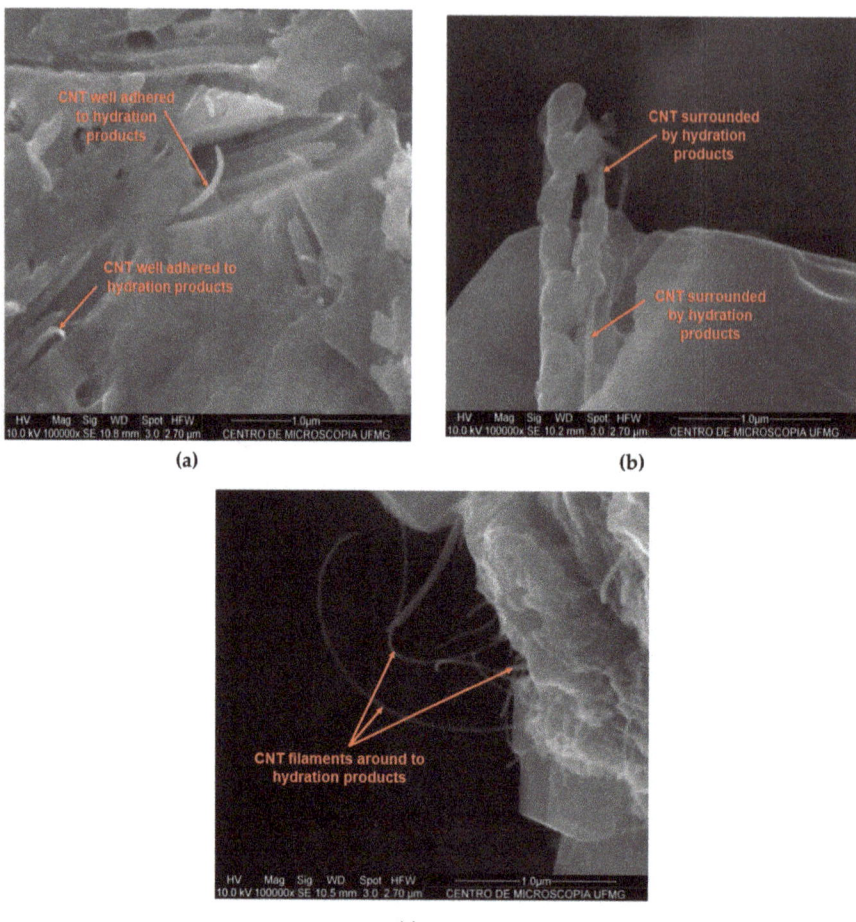

Figure 14. (a,b) Scanning electron microscopy images of hydrated cement pastes with 0.05% of CNTs (0.05-ISO); (c) Scanning electron microscopy images of hydrated cement pastes with 0.10% of CNTs (0.10-ISO).

In the images of cement paste fragments with 0.05% of CNTs (Figure 14a,b), it is difficult to identify CNTs. This fact could indicate that the CNTs were well dispersed and well incorporated in the cement hydration products. The hydration products could have hidden well-dispersed CNTs. The small filaments highlighted in Figure 14a can be identified as fragmented CNTs, evidencing good interaction with cement hydration products. Figure 14b displays a well-adhered CNT filament surrounded by hydration products, confirming that this nanomaterial acts as a nucleation site for cement hydration. In the 0.10-ISO paste, Figure 14c, CNTs are easier to visualize, which could indicate lower interaction with hydration products and worse dispersion as compared with 0.05-ISO.

In the proportion of 0.05% of CNTs, the maximum strength gain was recorded and the scanning electron microscopy images indicated better CNT dispersion. Otherwise, in the proportion of 0.10% of CNTs, although the strength gains recorded were compared to REF-ISO, there were indications that the effective dispersion limit was exceeded, resulting in agglomerations that possibly acted as strength concentration and contributed to lower strength as compared with 0.05-ISO.

4. Conclusions

The use of MWCNTs dispersed in non-aqueous isopropanol media on the surface of anhydrous cement particles was discovered to be an effective way for cement paste nanocomposite preparation. Nanostructured cement pastes in the proportions of 0.05% and 0.10% of CNTs in relation to the cement weight revealed higher compressive strength, flexural tensile strength, fracture energy, and fracture toughness. The gains obtained could evidence that CNTs act as nucleation of cement hydration products and fibrous reinforcement in cement-based material by this dispersion process.

The cement pastes with CNTs presented denser structure according to the water absorption test results, suggesting general pore refinement. The porosimetry analysis of nitrogen condensation in the mesoporous range indicated a lower quantity of finer pores (below 10 nm) in the presence of CNTs. These results are related to the CNTs' behavior as nucleation sites, supported by scanning electron microscopy images, where it is observed that the CNTs' filaments are surrounded and adhered to hydration products. Moreover, the thermogravimetric analysis also indicated that, in the presence of CNTs, the quantity of C-S-H and CH in the hydrated cement pastes was higher, corroborating with the statement that CNTs can additionally perform as nucleation sites of cement hydration products, resulting in a denser matrix with pore refinement. Additionally, the higher amount of carbonated material of REF paste indicates a reduction of permeability as an effect of CNT addition.

The comparison of the different concentrations demonstrated that the best result was achieved in the presence of 0.05% of CNTs. By the described methodology, the effective dispersion limit was reached in this proportion and in the ratio close to 0.89 g of CNTs per $m^2\ g^{-1}$ of anhydrous cement surface. The scanning electron microscopy images suggested that the amount of 0.10% of CNTs permitted the formation of small agglomerations that possibly contributed to lower strength regarding the paste with 0.05% CNT content.

Author Contributions: V.V.R. carried out the experiments and wrote the manuscript. P.L. conceived and planned the experiments. Both authors discussed the results and contributed to the final manuscript. All authors have read and agreed to the published version of the manuscript.

Funding: This research received funding from CAPES, CEFET-MG, CNPq (project number 459324/2014-4) and FAPEMIG (project number APQ-01087-14).

Acknowledgments: The authors would like to thank CAPES, CEFET-MG, CTNano, Center of Microscopy of UFMG, CNPq, and FAPEMIG for the materials made available and for the financial and technical assistance provided for this work.

Conflicts of Interest: The authors declare no conflict of interest.

References

1. Makar, J.M.; Beaudoin, J.J. Carbon nanotubes and their application in the construction industry. *Spec. Publ.-R. Soc. Chem.* **2004**. [CrossRef]
2. Collins, F.; Lambert, J.; Duan, W.H. The influences of admixtures on the dispersion, workability, and strength of carbon nanotube–OPC paste mixtures. *Cem. Concr. Compos.* **2012**. [CrossRef]
3. Hawreen, A.; Bogas, J.; Guedes, M.; Pereira, M.F.C. Dispersion and reinforcement efficiency of carbon nanotubes in cementitious composites. *Mag. Concr. Res.* **2018**. [CrossRef]
4. Rashad, A.M. Effect of carbon nanotubes (CNTs) on the properties of traditional cementitious Materials. *Constr. Build. Mater.* **2017**. [CrossRef]
5. Al-Rub, R.K.A.; Ashour, A.I.; Tyson, B.M. On the aspect ratio effect of multi-walled carbon nanotube reinforcements on the mechanical properties of cementitious nanocomposites. *Constr. Build. Mater.* **2012**. [CrossRef]
6. Hu, Y.; Luo, D.; Li, P.; Li, Q.; Sun, G. Fracture toughness enhancement of cement paste with multi-walled carbon nanotubes. *Constr. Build. Mater.* **2014**. [CrossRef]
7. Wang, B.; Han, Y.; Liu, S. Effect of highly dispersed carbon nanotubes on the flexural toughness of cement-based composites. *Constr. Build. Mater.* **2013**. [CrossRef]

8. Zou, B.; Chen, S.J.; Korayem, A.H.; Collins, F.; Wang, C.M.; Duan, W.H. Effect of ultrasonication energy on engineering properties of carbon nanotube reinforced cement pastes. *Carbon* **2015**. [CrossRef]
9. Rocha, V.V.; Ludvig, P. Nanocomposites prepared by a dispersion of CNTs on cement particles. *Archit. Civ. Eng. Environ.* **2018**, *11*. [CrossRef]
10. Li, G.Y.; Wang, P.M.; Zhao, X. Mechanical behavior and microstructure of cement composites incorporating surface-treated multi-walled carbon nanotubes. *Carbon* **2005**, *43*, 1239–1245. [CrossRef]
11. Hawreen, A.; Bogas, J.; Dias, A. On the mechanical and shrinkage behavior of cement mortars reinforced with carbon nanotubes. *Constr. Build. Mater.* **2018**. [CrossRef]
12. Rocha, V.V.; Ludvig, P.; Trindade, A.C.C.; de Andrade Silva, F. The influence of carbon nanotubes on the fracture energy, flexural and tensile behavior of cement based composites. *Constr. Build. Mater.* **2019**, *209*, 1–8. [CrossRef]
13. Carriço, A.; Bogas, J.A.; Hawreen, A.; Guedes, M. Durability of multi-walled carbon nanotube reinforced concrete. *Constr. Build. Mater.* **2018**, *164*, 121–133. [CrossRef]
14. Nochaiya, T.; Chaipanich, A. Behavior of multi-walled carbon nanotubes on the porosity and microstructure of cement-based materials. *Appl. Surf. Sci.* **2011**, *257*, 1941–1945. [CrossRef]
15. Xu, S.; Liu, J.; Li, Q. Mechanical properties and microstructure of multi-walled carbon nanotube-reinforced cement paste. *Constr. Build. Mater.* **2015**, *76*, 16–23. [CrossRef]
16. Isfahani, F.T.; Li, W.; Redaelli, E. Dispersion of multi-walled carbon nanotubes and its effects on the properties of cement composites. *Cem. Concr. Compos.* **2016**, *74*, 154–163. [CrossRef]
17. Souza Filho, A.G.; Fagan, S.B. Funcionalização de nanotubos de carbono. *Quím. Nova* **2007**. [CrossRef]
18. Liu, Y.; Gao, L.; Sun, J. Noncovalent functionalization of carbon nanotubes with sodium lignosulfonate and subsequent quantum dot decoration. *J. Phys. Chem. C* **2007**. [CrossRef]
19. Batiston, E.R.; Hampinelli, D.; Oliveira, R.C.; Gleize, P.J.P. Funcionalização e efeito da incorporação de nano tubos de carbono na cinética de hidratação em matrizes cimentícias. *Congr. Bras. Concr.* **2010**, *52*, 1–12.
20. Ludvig, P.; Calixto, J.M.; Ladeira, L.O.; Gaspar, I.C. Using converter dust to produce low cost cementitious composites by in situ carbon nanotube and nanofiber synthesis. *Materials* **2011**, *4*. [CrossRef]
21. Alsharefa, J.M.; Tahaa, M.R.; Khan, T.A. Physical dispersion of nanocarbons in composites—A review. *Technol. J.* **2017**. [CrossRef]
22. Makar, J.M.; Chan, G.W. Growth of cement hydration products on single-walled carbon nanotubes. *J. Am. Ceram. Soc.* **2009**. [CrossRef]
23. Makar, J.; Margeson, J.; Luh, J. Carbon nanotube/cement composites-early results and potential applications. In Proceedings of the 3rd International Conference on Construction Materials: Performance, Innovation and Structural Implications, Vancouver, BC, Canada, 21 August 2005; pp. 1–10.
24. Silva, R.A.; Guetti, P.; Da Luz, M.S.; Rouxinol, F.; Gelamo, R.V. Enhanced properties of cement mortars with multilayer graphene nanoparticles. *Constr. Build. Mater.* **2017**. [CrossRef]
25. Dally, J.W.; Riley, W.F. *Experimental Stress Analysis*, 3rd ed.; McGraw-Hill: New York, NY, USA, 1991; p. 639.
26. American Society for Testing Materials. *ASTM 349-02. Standard Test Methods for Compressive Strength of Hydraulic-Cement Mortars (Using Portions of Prisms Broken in Flexure)*; American Society for Testing Materials: West Conshohocken, PA, USA, 2002. [CrossRef]
27. ASTM International. *ASTM C642-13. Standard Test Method for Density, Absorption, and Voids in Hardened Concrete*; ASTM International: West Conshohocken, PA, USA, 2013. [CrossRef]
28. Korpa, A.; Trettin, R. The influence of different drying methods on cement paste microstructures as reflected by gas adsorption: Comparison between freeze-drying (F-drying), D-drying, P-drying and oven-drying methods. *Cem. Concr. Res.* **2006**. [CrossRef]
29. Balbo, J.T. Relações entre resistências à tração indireta e à tração na flexão em concretos secos e plásticos. *Rev. IBRACON Estrut. Mater.* **2013**, *6*, 6. [CrossRef]
30. Bažant, Z.P. Size effect. *Int. J. Solids Struct.* **2000**, *37*, 69–80. [CrossRef]
31. Hu, X.; Duan, K. Size effect and quasi-brittle fracture: The role of FPZ. *Int. J. Fract.* **2008**, *154*, 3–14. [CrossRef]
32. Ludvig, P.; Calixto, J.M.F.; Ladeira, L.O.; Souza, T.C.; Paula, J.N. Analysis of Cementitious Composites Prepared with Carbon Nanotubes and Nanofibers Synthesized Directly on Clinker and Silica Fume. *J. Mater. Civ. Eng.* **2017**, *29*. [CrossRef]
33. Sing, K.S.; Williams, R.T. Physisorption hysteresis loops and the characterization of nanoporous materials. *Adsorpt. Sci. Technol.* **2004**, *22*, 773–782. [CrossRef]

34. Fordham, C.J.; Smalley, I.J. A simple thermogravimetric study of hydrated cement. *Cem. Concr. Res.* **1985**, *15*, 141–144. [CrossRef]
35. Almeida, A.E.F.D.S.; Tonoli, G.H.D.; Santos, S.F.D.; Savastano, H. Improved durability of vegetable fiber reinforced cement composite subject to accelerated carbonation at early age. *Cem. Concr. Compos.* **2013**, *42*, 49–58. [CrossRef]
36. Ma, Q.; Guo, R.; Zhao, Z.; Lin, Z.; He, K. Mechanical properties of concrete at high temperature—A review. *Constr. Build. Mater.* **2015**, *93*, 371–383. [CrossRef]
37. Bonnaud, P.A.; Ji, Q.; Coasne, B.; Pellenq, R.M.; Van Vilet, K.J. Thermodynamics of water confined in porous calcium-silicate-hydrates. *Langmuir* **2012**, *28*, 11422–11432. [CrossRef] [PubMed]

 © 2020 by the authors. Licensee MDPI, Basel, Switzerland. This article is an open access article distributed under the terms and conditions of the Creative Commons Attribution (CC BY) license (http://creativecommons.org/licenses/by/4.0/).

Article

Comparing Multi-Walled Carbon Nanotubes and Halloysite Nanotubes as Reinforcements in EVA Nanocomposites

Agata Zubkiewicz [1],*, Anna Szymczyk [1], Piotr Franciszczak [2], Agnieszka Kochmanska [2], Izabela Janowska [3] and Sandra Paszkiewicz [2],*

1. Department of Technical Physics, West Pomeranian University of Technology, 70311 Szczecin, Poland; aszymczyk@zut.edu.pl
2. Department of Materials Technology, West Pomeranian University of Technology, 70310 Szczecin, Poland; piotr.franciszczak@zut.edu.pl (P.F.); akochmanska@zut.edu.pl (A.K.)
3. Institut de Chimie et Procédés pour l'Energie l'Environnement et la Santé (ICPEES), University of Strasbourg, 67087 Strasbourg, France; janowskai@unistra.fr
* Correspondence: agata.zubkiewicz@zut.edu.pl (A.Z.); spaszkiewicz@zut.edu.pl (S.P.); Tel.: +48-91-449-4589 (S.P.)

Received: 19 July 2020; Accepted: 25 August 2020; Published: 28 August 2020

Abstract: The influence of carbon multi-walled nanotubes (MWCNTs) and halloysite nanotubes (HNTs) on the physical, thermal, mechanical, and electrical properties of EVA (ethylene vinyl acetate) copolymer was investigated. EVA-based nanocomposites containing MWCNTs or HNTs, as well as hybrid nanocomposites containing both nanofillers were prepared by melt blending. Scanning electron microcopy (SEM) images revealed the presence of good dispersion of both kinds of nanotubes throughout the EVA matrix. The incorporation of nanotubes into the EVA copolymer matrix did not significantly affect the crystallization behavior of the polymer. The tensile strength of EVA-based nanocomposites increased along with the increasing CNTs (carbon nanotubes) content (increased up to approximately 40% at the loading of 8 wt.%). In turn, HNTs increased to a great extent the strain at break. Mechanical cyclic tensile tests demonstrated that nanocomposites with hybrid reinforcement exhibit interesting strengthening behavior. The synergistic effect of hybrid nanofillers on the modulus at 100% and 200% elongation was visible. Moreover, along with the increase of MWCNTs content in EVA/CNTs nanocomposites, an enhancement in electrical conductivity was observed.

Keywords: EVA elastomers; halloysite nanotubes (HNTs); carbon nanotubes (CNTs); nanocomposites; thermal properties; mechanical properties; electrical conductivity

1. Introduction

In recent years, polymer nanocomposites containing various nanofillers such as graphite nanoplatelets and carbon nanotubes as well as montmorillonite nanoclays have attracted enormous interest in academic and industrial fields. the unique properties of the polymer nanocomposites, such as flame retardancy, improved thermal stability, increased mechanical properties, and gas barrier properties, depend not only on the properties of nanofillers and polymer matrix but also on the interfacial contact and interactions between the nanofiller and the polymer matrix.

Among the vast range of nanofillers, one of the most promising are carbon nanotubes (CNTs). They were discovered in 1991 by Sumio Iijima [1]. CNTs can be broadly categorized as single-walled carbon nanotubes (SWCNTs), with a typical diameter between 1 and 2 nm, and multi-walled carbon nanotubes (MWCNTs), with an outer diameter between 3 and 30 nm or more, depending on the number of graphitic layers forming their structure [2]. They exhibit excellent mechanical (elastic modulus:

1–1.7 TPa), thermal (thermal conductivity higher than 3000 W·m^{-1}·K^{-1}), and electrical (electrical conductivity: 10^5 S·m^{-1}–10^7·S·m^{-1}) properties [3]. Moreover, CNTs possess low mass density, and large aspect ratio (length to diameter ≈ 1000) [4]. Despite their numerous advantages, carbon nanotubes also suffer from a few drawbacks. Because of the nanometric dimensions, CNTs have a strong tendency to aggregate. Moreover, carbon nanotubes are relatively expensive. Despite this, CNTs have gained a great deal of research interest, for over 20 years, especially as a reinforcing nanofillers in polymer nanocomposites [5]. In particular, great attention has been focused on multiwalled carbon nanotubes (MWCNTs) as fillers in polymer materials, such as epoxy resins [6–8], polyethylene (PE) [9,10], polypropylene (PP) [11–13], polyurethanes (PU) [14,15], etc.

Another type of nanotubular structures is naturally occurring halloysite nanotubes (HNTs). They were reported for the first time in 1826 by Berthier [16]. HNTs are abundantly available nanoparticles formed by rolled kaolin sheets with chemical composition $Al_2Si_2O_5(OH)_4·2H_2O$ (the hydrated form with one layer of water in the interlayer spaces: HNTs-10 Å). In a dry climate, they can also occur in an anhydrous form (with an interlayer spacing of 7Å) with the formula $Al_2Si_2O_5(OH)_4$ [17]. HNTs are 1:1 phyllosilicates that have a hollow tubular morphology which results from the wrapping of silicate sheets, consisting of one tetrahedral and one octahedral sheet, that are connected through hydrogen bonding and weak Van der Waals interactions [18]. HNTs were found to occur in soils all over the world, but the most important deposits are located in the United States, New Zealand, and Poland [19]. the pure material is white, however, as a result of impurities from ferric ions, it may be slightly red [20]. Typically, the HNTs lengths range between 300 and 1500 nm, while their inner diameters are 15–100 nm, and outer diameter 40–120 nm [21,22]. HNTs are low-cost, and eco-friendly materials that can be more easily dispersed in a polymer matrix than carbon nanotubes [19]. Compared to other layered silicates, they are characterized by relatively low hydroxyl content on their outer surfaces. It is related to the fact that most of the aluminols are located inside the tubes. In the outer surface of the HNTs are located mainly siloxane groups [16,23]. Moreover, HNTs have quite a high aspect ratio (10–50), high resistance to heat and chemical substances, and due to the empty lumen structures, relatively low density (2.14–2.59 g/cm^3) [19]. The surface of HNTs is negatively charged (at pH > ~2), whereas the inner surface is charged positively [16]. Because of the fact that HNTs are naturally occurring and much cheaper, yet morphologically similar to multiwalled carbon nanotubes, the HNTs could be an alternative for more expensive CNTs for selected applications. A lot of research has been carried out on the nanocomposites based on HNTs with various polymer such as PE [24,25], PP [23,26,27], PA [22] epoxy resin [28,29]. They can provide a significant improvement in the thermal stability, fire resistance, and mechanical properties of composites.

A range of novel materials can be obtained by the simultaneous introduction of two types of nanofillers to the polymer matrix. Hybrid materials combine the properties of both fillers, and may also exhibit additional properties, because of the synergistic effects [30–34]. Recent research on the manufacturing of hybrid thermoplastic composites focuses on hybrid reinforcement in order to achieve better or tailored mechanical performance [35–39]. In the case of nano-reinforcements, this usually pertains to obtaining some distinctive physical properties e.g., electrical conductivity, thermal resistance, or barrier properties.

Ethylene vinyl acetate (EVA) is an important copolymer, widely used in various applications such as wire and cable insulations, shoe soles, noncorrosive layers, and component packaging. EVA has good flexibility, low cost, and good barrier properties. On the other hand, EVA has low tensile strength, thermal stability, and high flammability. All these disadvantages can be overcome by adding nanofillers to the polymer matrix. There are relatively fewer works describing the effect of HNTs' content on the mechanical and thermal properties of EVA. Suvendu Padhi et al. reported that HNTs could improve the mechanical properties and thermal stability of EVA [18]. Moreover, the addition of HNTs to EVA could enhance water resistance and oxygen permeability [40]. Various CNTs containing EVA-based nanocomposites had also been reported [41–45]. The mass production of high quality CNTs at lower cost and their exceptional electrical, thermal, and mechanical properties, make it one of the most attractive nanofillers. The addition of carbon nanotubes to the EVA matrix can improve mechanical, thermal, and electrical characteristics.

In this work, we manufactured and compared nanocomposites based on EVA copolymer containing MWCNTs, HNTs, and the mixture of both of them (hybrid system) in mass ratio 1:1. The morphology, thermal, mechanical, and electrical properties of the manufactured nanocomposites were characterized.

2. Materials and Methods

2.1. Materials

Ethylene vinyl acetate copolymer (EVA Elvax 40L-03, DuPont DuPont Company, Wilmington, DE, USA) containing 40 wt.% of vinyl acetate was applied as a matrix in the obtained nanocomposites. According to producer data, it has a density of 0.967 g/cm^3 and a melt flow rate of 3 g/10 min (at 190 °C and 2.16 kg). Halloysite nanotubes (HNTs) with a diameter of 30–70 nm, length of 1–3 µm, density of 2.53 g/cm^3, the pore size of 1.26–1.34 mL/g pore volume, and surface area of 64 m^2/g were obtained from Sigma-Aldrich. Multi-walled carbon nanotubes (MWCNTs, Nanocyl®® NC7000^{TM}) were purchased from Nanocyl SA (Sambreville, Belgium). The nanotubes had an average diameter of 9.5 nm and a length of 1.5 µm, and the specific surface area of 250–300 m^2/g, the density of 1.75 g/cm^3, volume resistivity of 10^{-4} Ω·cm, according to the supplier's specification.

2.2. Composite Manufacturing

2.2.1. Compounding

The nanocomposites based on EVA containing MWCNTs, HNTs, or the hybrid system of MWCNTs/HNTs (1:1) were prepared by melt blending using a counter-rotating, tight intermeshing twin-screw extruder: Leistritz Laborextruder LSM30 (L/D = 23, D = 34 mm). The nanofillers and EVA granulate were fed separately using two gravimetric feeders into their feed section. The compounding was carried out at temperatures ranging from 50 to 115 °C from the feed section to the nozzle and at 40 rpm. The extruded strands of compounds were then cooled in a water tank and subsequently pelletized. Three series of nanocomposites with different content of nanofillers were prepared: nanocomposites containing MWCNTs with filler content of 2, 4, 6, and 8 wt.%, nanocomposites containing HNTs with filler content of 2, 4, 6, and 8 wt.%, and hybrid nanocomposites containing both MWCNTs and HNTs (at mass ratio 1:1) with the total fillers' content of 4 and 6 wt.%. EVA compounds with MWCNTs, so as with the mixture of MWCNTs with HNTs, were manufactured in one-step compounding. The compounds with HNTs in turn were manufactured in two-step compounding due to hindrance in the feeding of small portions of HNT nanoclay, which has high bulk density. For this purpose, the masterbatch of EVA/HNT 68/32 wt.% was manufactured and was subsequently diluted with EVA to the set filing ratios in the second compounding process. The set and obtained filling ratios of manufactured EVA compounds, as well as their corresponding volumetric filling ratios are presented in Table 1.

2.2.2. Injection Molding

Nanocomposites were dried before injection molding at 43 °C for ~12 h in a POL-EKO SLW115 oven (POL-EKO-APARATURA sp.j., Wodzislaw Slaski, Poland) with forced convection. The standard test specimens were injection molded using an ARBURG ALLROUNDER (ARBURG GmbH + Co. KG, Lossburg, Germany) 270 S 350–100 (clamping force 350 kN, screw diameter 25 mm, L/D = 20). Type A (dogbone) specimens for tensile testing were manufactured in accordance with EN ISO 294, while samples of dimensions 60 × 60 × 2 mm were manufactured for electrical conductivity measurements. The barrel temperatures were set to 100–150 °C from the first zone to the nozzle. Mold temperature was kept ~30 °C. In order to bring the processing of samples closer to conditions of extrusion, a relatively slow constant injection volume flow of 10 ccm/s was used during injection molding, which resulted in actual injection pressures presented in Table 1. Injection time was ~3.7 s, while the holding pressure was set to rise from 400 to 1000 bars for 15 s For EVA/HNT compounds the holding pressure was reduced to 800 bars. Backpressure by dosing was set to 30 bars. Cooling time was 30 s The whole injection molding cycle amounted to around minute. In turn, MWCNT's increased the melt viscosity, which was reflected by up to 50% higher injection molding pressures for the highest 8 wt.% filling ratio.

Table 1. Compositions and the injection pressure of ethylene vinyl acetate (EVA)-based nanocomposites.

Material	Set Filler Content (wt.%)	Filler Real Content (wt.%)	MWCNTs Volumetric Content (vol.%)	HNTs Volumetric Content (vol.%)	Injection Pressure-Type A Samples (bars)	Injection Pressure-Conductivity Samples (bars)
EVA	-	-	-	-	800	640
EVA/2 wt.% CNT	2	2.16	1.18	-	820	660
EVA/4 wt.% CNT	4	3.93	2.16	-	910	710
EVA/6 wt.% CNT	6	5.8	3.22	-	1040	910
EVA/8 wt.% CNT	8	7.94	4.45	-	1170	970
EVA/2 wt.% HNT	2	2.01	-	0.78	690	560
EVA/4 wt.% HNT	4	4.02	-	1.57	680	550
EVA/6 wt.% HNT	6	6.03	-	2.38	710	590
EVA/8 wt.% HNT	8	8.03	-	3.20	730	600
EVA/4 wt.% CNT + HNT (1:1)	4	3.97	1.09	0.78	810	660
EVA/6 wt.% CNT + HNT (1:1)	6	6.05	1.68	1.20	900	710

2.3. Measurements

The morphology of the nanocomposites was analyzed using a scanning electron microscope (FE-SEM, Hitachi SU-70, Naka, Japan). Before SEM analysis, the samples were cryofractured in liquid nitrogen and then coated with a thin film of palladium-gold alloy, using thermal evaporation PVD (physical vapor deposition) method to provide electric conductivity.

Transmission electron microscopy (TEM) was performed on a 2100 F Jeol microscope. Prior to the analysis, the samples were cooled down in liquid nitrogen and cut into thin-layered pieces.

The X-ray Diffraction (XRD) analysis of the EVA-based nanocomposites was conducted with the use of a Panalytical X'Pert diffractometer (Malvern Panalytical, Malvern, UK) operating at 40 V and 40 mA with CuKα radiation (λ = 0.154 nm). The samples were scanned from 2θ = 4° to 70° with a step of 0.02°.

The Fourier Transform Infrared Spectroscopy (FTIR) spectra of the EVA-based nanocomposites were recorded on a Tensor-27 spectrophotometer (Brucker, Ettlingen, Germany), in the range of 400–4000 cm^{-1}. Measurements were carried out using the attenuated total reflectance (ATR) technique.

Differential scanning calorimetry (DSC) measurements of the obtained nanocomposites were conducted on a Mettler Toledo DSC1 instrument (Mettler Toledo GmbH, Greifensee, Switzerland). The samples were heated up/cooled down under nitrogen flow with a heating rate of 10 °C/min in the temperature range −75 °C to 175 °C. The crystallinity degree (X_c) of the investigated materials was calculated using the following equation:

$$X_c = \Delta H_m / \Delta H_m^o (1 - \varphi_n) \qquad (1)$$

where: ΔH_m^o (= 293 J/g) is the enthalpy change of melting for a fully crystalline PE [46], φ_n is a weight content of nanofiller, and ΔH_m is derived from melting peak area on DSC thermogram.

Thermal and thermo-oxidative stability of prepared nanocomposites was determined by thermogravimetric analysis (TGA 92-16, Setaram, Caluire-et-Cuire, France). Samples were heated in nitrogen and oxidizing atmosphere ($N_2:O_2$ = 80:20 vol.%) from temperature range 20 °C to 700 °C.

Density (d_R) was measured at 22 °C, using hydrostatic scales (Radwag WPE 600C, Radom, Poland), calibrated using working standards of known density. For each sample, five measurements were conducted, and then the results were averaged to obtain a mean value. Theoretical densities of the obtained materials were calculated using the rule of mixture, taking into account the real density of EVA, filler densities according to the producer data, and their volume fractions.

The melt flow rate (MFR) was measured on a melt indexer (CEAST, Pianezza TO, Italy) at a temperature of 190 °C and under 2.19 kg load, according to ISO 1133 specification.

Hardness was measured using the Shore D apparatus (Zwick GmbH, Ulm, Germany) after 15 s of loading, according to a standard ISO 868. Ten measurements were conducted and then the results were averaged to obtain a mean value.

The tensile properties of the prepared nanocomposites were measured using Autograph AG-X plus universal testing machine (Shimadzu, Duisburg, Germany) equipped with a 1 kN Shimadzu load cell, a non-contact optical extensometer, and the TRAPEZIUM X computer software (version 1.00 provided by Shimadzu, Duisburg, Germany). Tests were performed at room temperature, with a strain rate of 250 mm/min to break. Tensile strength and elongation at break of the nanocomposites were determined. The reported values are the average values of ten measurements. Cyclic tensile measurements were performed using the same equipment, with a testing speed of 100 mm/min. The samples were stretched until the specified strain value was reached and then the tensile force was released to the zero. This procedure was repeated, with increasing deformation value, until the sample broke. The following strains were established for our test: 5%, 15%, 25% 50% 100%, 200%, and 400%.

The electrical conductivity of the obtained nanocomposites was evaluated by measurements of resistivity using Electrometer 6517A (Keithley Instruments, Inc., Germering, Germany) device together with a set of Keithley 8009. The measurements were performed accordingly to the standard PN-88/E-04405.

3. Results and Discussion

3.1. Dispersion State Investigation

The quality of the dispersion of filler in a matrix is a crucial factor that determines the final properties of composite material. The nanofillers dispersion was examined through scanning electron microscopy (SEM) and transmission electron microscopy (TEM). The representative micrographs of the nanocomposites containing 4 and 6 wt.% of nanofillers are shown in Figure 1. MWCNTs distribution appears to be rather homogenous in the entire polymer matrix, and no large agglomerates were observed (Figure 1a,b). However, from the comparison of EVA/CNTs and EVA/HNTs images, it is clear, that HNTs are more evenly distributed. CNTs tend to bundle together because of the van der Waals interaction between the individual nanotubes. In contrast to carbon nanotubes, HNTs have relatively low tube–tube interactions. It is related to the hydroxyl groups that are located on the HNTs surfaces. Moreover, the tubular morphologies with a relatively high aspect ratio limit the possibility of creating large-area contact between tubes [19]. Consequently, a very uniform dispersion of HNTs in the EVA matrix was obtained (Figure 1c,d). The SEM images of hybrid nanocomposites are presented in Figure 1e,f. Both nanofillers are well distributed within the polymer matrix. On the cryo-fractured surface of nanocomposites, the nanotubes pulled from the EVA copolymer matrix can be observed.

To confirm the state of dispersion of nanotubes in the matrix, TEM analysis for nanocomposite with the highest content of the MWCNTs and MWCNT/HNT hybrid system was also carried out. Figure 2 presents the TEM images of nanocomposites containing MWCNTs. Figure 3 and Figure S1 (Supplementary Materials) represent the TEM images of EVA/6 wt.% CNT + HNT hybrid nanocomposite at different magnification. In Figure 3a, it can be seen that the dispersed halloysite nanotubes are relatively short. Probably, some of HNTs broke during compounding. Moreover, in contrast to the SEM observation, on TEM images a few agglomerates of nanotubes can be seen. At the higher magnification on TEM images (Figure 2 and Figure S2), it can be seen that the individual and entangled agglomerates of CNTs were embedded in the EVA matrix.

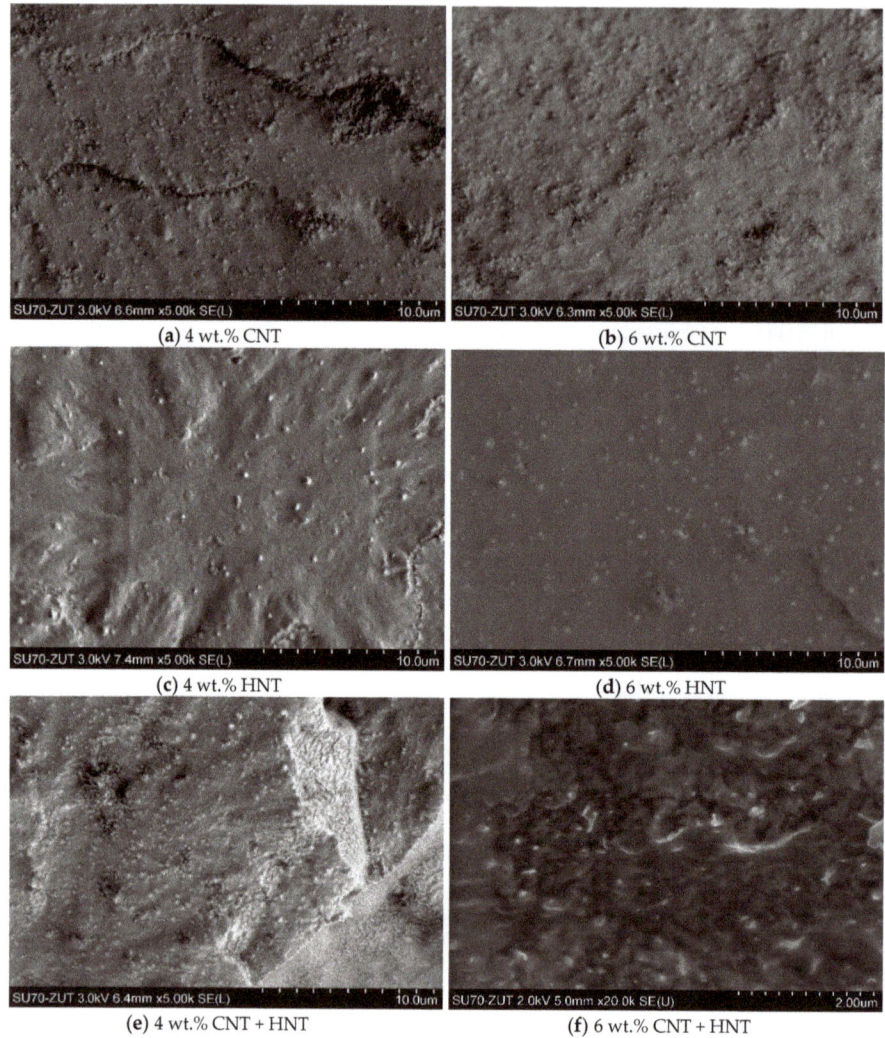

Figure 1. SEM images of EVA/multi-walled carbon nanotubes (MWCNT) (**a**,**b**), EVA/halloysite nanotubes (HNT) (**c**,**d**), and EVA/MWCNT + HNT (**e**,**f**) nanocomposites.

3.2. XRD Analysis

The XRD patterns of the pure HNTs, MWCNTs powder, and EVA nanocomposites with their different loading are shown in Figure 4. The pristine HNTs shows a characteristic reflection peaks at diffraction angles (2θ) of 11.71, 19.99, 24.88, 26.63, 35.1 35.92, 38.31, and 62.40° corresponding to the d-spacings of 0.755 nm (001), 0.443 nm (020), 0.357 nm (110), 0.334 nm (113), 0.256 nm, (131), 0.234 nm (131)), 0.234 nm (133), and 0.148 nm (332), respectively [47]. The diffraction peaks at around 2θ = 12° and 2θ = 25° are attributed to the dehydrated form of HNTs, whereas the visible distinct peak at around 2θ = 63° indicates that the halloysite is a dioctahedral mineral [18,40]. For neat EVA, only a broad scattering reflection is found. It is located at around 2θ = 20.4° and indicates mainly an amorphous structure of the neat EVA matrix because of the low crystallinity of neat EVA (Table 2). It can be seen, that for EVA/HNTs nanocomposites, the HNTs characteristic peaks at 2θ of 12°

and 25° are visible (Figure 4a) and thus, with the increase in the addition of HNTs, the intensity of these peaks increases. The position of diffraction peaks of EVA-based nanocomposites remained unchanged with different HNTs content. Figure 4b shows the XRD patterns of CNT, EVA/CNT, and hybrid nanocomposites. The characteristic peaks assigned to MWCNTs are seen at 2θ of 25.5° (d = 0.348 nm) and 42.7° (d = 0.211 nm), and they correspond respectively to the graphite indices of (002), which is related to d-spacing between graphene sheets, and (100), which is associated to the in-plane graphitic structure [48,49]. These characteristic peaks for MWCNTs were not present in XDR patterns for EVA/CNTs nanocomposites, because of the overlapping diffraction signals of CNTs and EVA and their low intensity (low MWCNTs concentration) in EVA matrix. The XRD patterns of EVA hybrid nanocomposites show a characteristic peak of HNTs at 2θ = 12.2°. The other diffraction peaks overlap with the peaks of EVA copolymer.

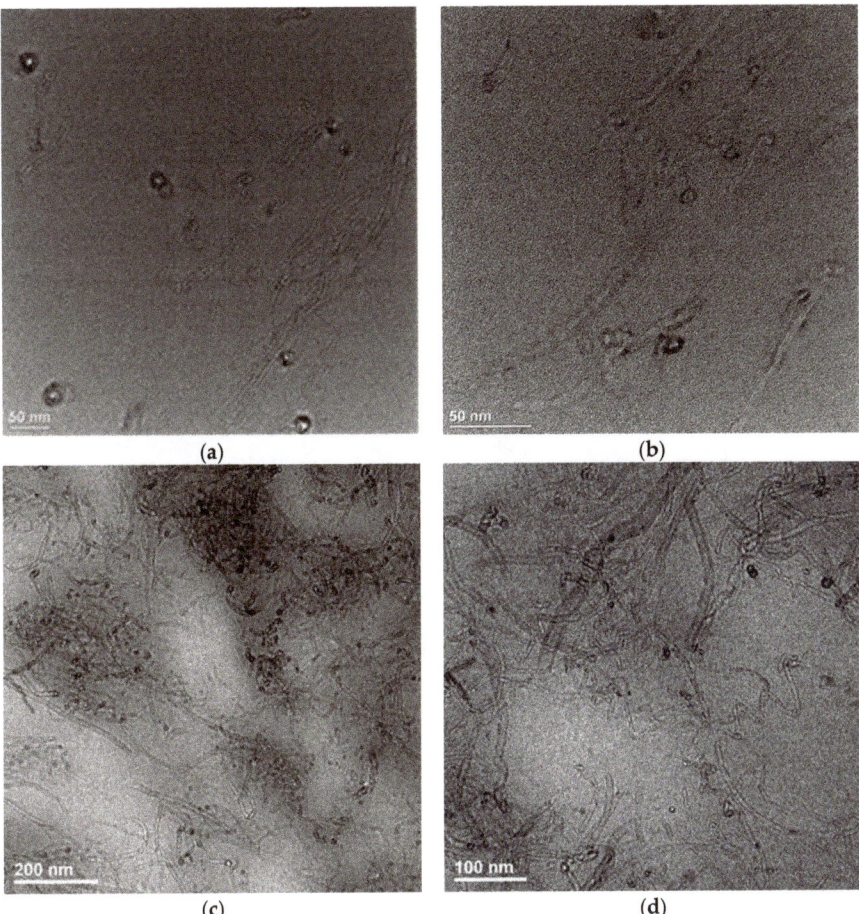

Figure 2. TEM images of EVA/6 wt.% CNT (**a,b**), EVA/8 wt.% CNT (**c,d**) nanocomposites at different magnification. Scale bars: 50 nm (**a,b**), 200 nm (**c**), 100 nm (**d**).

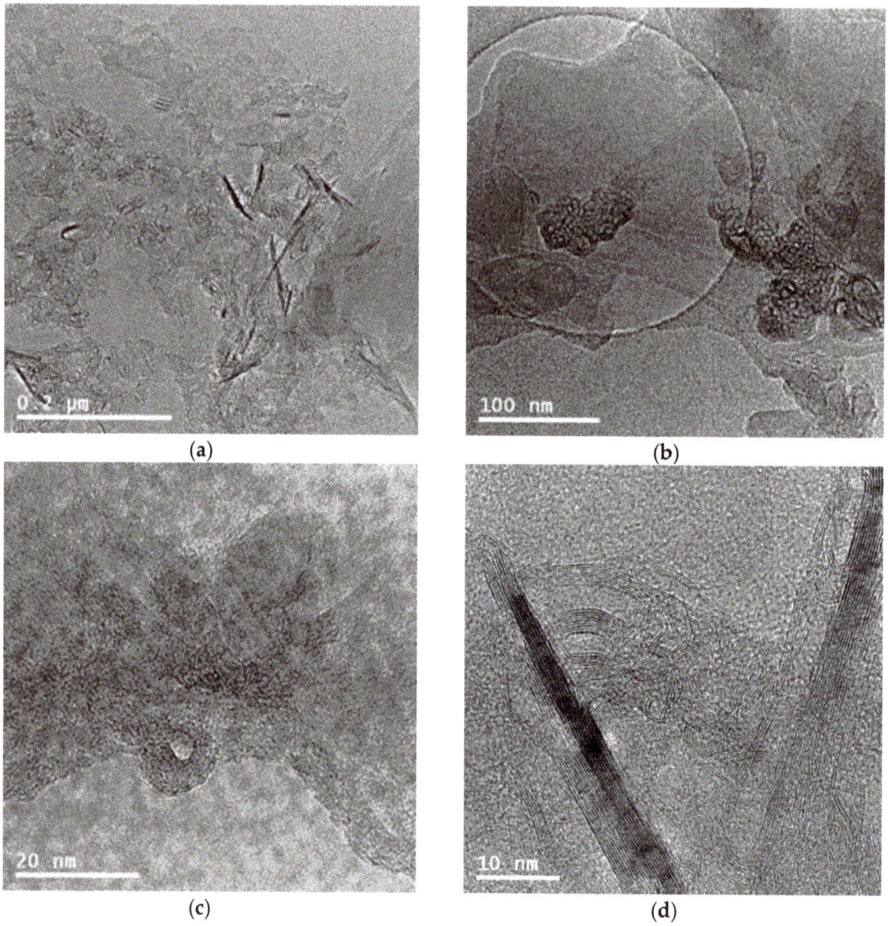

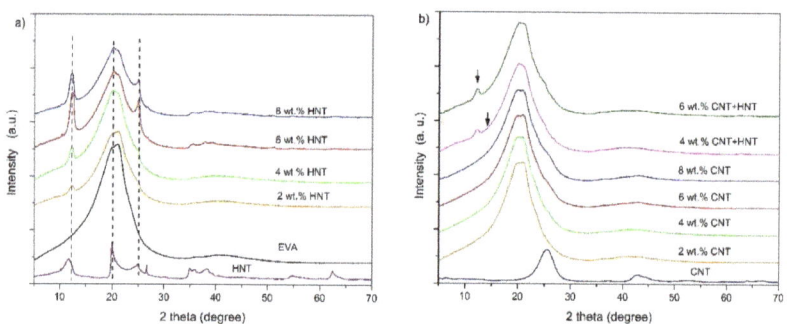

Figure 3. TEM images of EVA/6 wt.% CNT + HNT hybrid nanocomposite at different magnification. Scale bars: 0.2 μm (**a**), 100 nm (**b**), 20 nm (**c**), 10 nm (**d**).

Figure 4. X-ray diffraction patterns of EVA/HNTs (**a**), EVA/MWCNTs, and MWCNT/HNT hybrid (**b**) nanocomposites.

Table 2. Phase transition temperatures, enthalpies of melting, and degree of crystallinity of EVA-based nanocomposites.

Material	T_g °C	T_m °C	H_m J/g	T_c °C	X_c%
EVA	−20	50	5.79	26	1.98
EVA/2 wt.% CNT	−18	51	5.88	24	2.00
EVA/4 wt.% CNT	−17	51	5.29	23	1.81
EVA/6 wt.% CNT	−16	50	5.39	23	1.84
EVA/8 wt.% CNT	−14	50	5.06	22	1.73
EVA/2 wt.% HNT	−18	51	5.78	26	1.97
EVA/4 wt.% HNT	−18	51	5.35	26	1.83
EVA/6 wt.% HNT	−18	51	5.32	26	1.82
EVA/8 wt.% HNT	−17	51	4.60	26	1.57
EVA/4 wt.% CNT + HNT	−17	52	5.37	23	1.83
EVA/6 wt.% CNT + HNT	−17	51	5.61	23	1.92

T_g—glass transition temperature; T_m—melting temperature; ΔH_m—enthalpy of melting; T_c—crystallization temperature; X_c—degree of crystallinity.

3.3. Thermal Properties of the EVA Nanocomposites

The DSC thermograms for EVA/CNT nanocomposites recorded during first (dashed line) and second heating, as well as cooling, as plotted in Figure 5a,b, respectively. Similarly, the DSC thermograms for the series of materials containing HNTs are presented in Figure 5c,d. Besides, in Figure 5e,f, the DSC thermograms were plotted for the samples containing 4 wt.% (Figure 5e) and 6 wt.% (Figure 5f) (in total) of CNT, HNT, and the mixture of both (CNT + HNT) at a mass ratio of 1:1. Likewise, the DSC parameters are summarized in Table 2. For the series of EVA-based nanocomposites containing CNTs, one observed that along with an increase of the CNTs' concentration, the value of the glass transition temperature (T_g) of EVA also increases. The melting temperatures (T_m) of nanocomposites in comparison to the neat EVA copolymer were comparable to one another, while a slight decrease in the crystallization temperature (T_c) was observed. EVA copolymer containing 40% by mass of VA has a low crystallinity of around 2%. The addition of MWCNTs and HNTs did not significantly affect the degree of crystallinity of nanocomposites, their degree of crystallinity is around 1.6–2.0%. Generally, CNTs have been proved to be good nucleating agents for polymer crystallization [50,51]. It has been reported that at low loading of CNTs in some semicrystalline polymers in the molten state they can induce crystallization at higher temperatures through decreasing the nucleation activation energy and increasing the nucleation density, leading to the acceleration of crystallization and the decrease of spherulites diameters simultaneously [52,53]. It was also found that the CNTs in some polymer systems can generate anti-nucleation effects, and super-nucleation effects on polymer matrices [54,55]. No effect of CNTs on the nucleation of polymer crystals has also been reported in some cases [56,57]. More recently, LDPE/CNTs nanocomposites have been prepared in our research group [58]. Our results show that crystallization behavior of PE in LDPE/CNT nanocomposites was not influenced by the presence of CNTs. However, herein in EVA nanocomposites, rather antinucleating behavior of CNTs was observed. As can be seen in Table 2, the crystallization temperature (T_c) of EVA nanocomposites decreases with the increase of CNTs content. Similar as in LDPE/CNTs nanocomposites, it can be a result of a low value of surface energy and a poor wettability of CNTs. This means that it can be difficult for them to induce aggregation of polymer chains on their surfaces [58]. In turn, the increase in T_g results from the fact that CNTs may prevent the mobility of the copolymer chains, leading to an increase in T_g [50]. For the series of EVA-based nanocomposites containing HNTs, similarly as in the case of the CNTs, the values of T_m and T_c for nanocomposites are comparable to the value observed for neat EVA copolymer. However, along with an increase in the concentration of HNTs, the value of T_g increased. In some polymer systems, halloysite nanotubes act as nucleating agents accelerating crystallization rate [26]. The introduction of HNTs into EVA copolymer did not influence T_c and the degree of nanocomposites crystallinity in comparison to the neat EVA copolymer. Therefore, one can deduce that HNTs rather frustrates chain ordering and mobility, than the crystallinity behavior itself [40]. In turn, for two hybrid nanocomposites containing 4 wt.% and 6 wt.% of CNTs and HNTs, the observations on the phase transition temperatures are comparable to the ones observed for EVA/CNTs and EVA/HNTs nanocomposites, i.e., comparable values of T_m, slightly lower values of T_c and X_c, and increase in T_g. These values appear to be the result of the impact of both types of nanoparticles, without the apparent effect of any of the used ones.

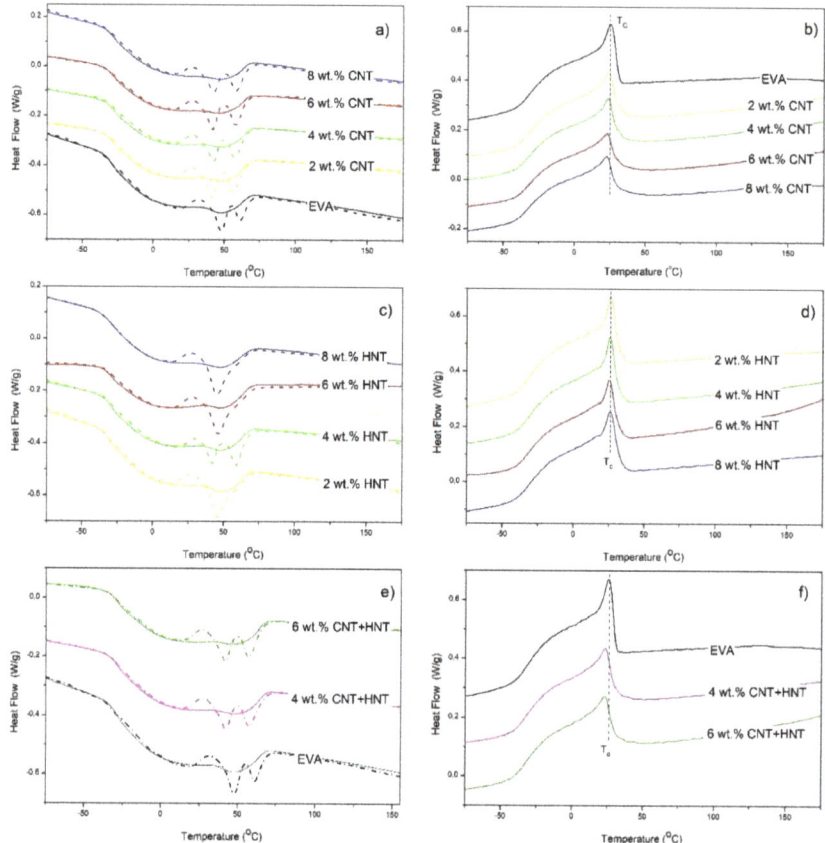

Figure 5. DSC thermograms for EVA/MWCNT nanocomposites (**a**,**b**), EVA/HNT nanocomposites (**c**,**d**), and hybrid nanocomposites (**e**,**f**).

Additionally, since the addition of carbon nanofillers, like CNTs, mineral nanotubes like HNTs or mineral nanoclays like MMT, can improve thermal and thermo-oxidative stability of polymer matrices, the mass loss and derivative mass loss curves for the series of materials have been presented in Figures 6 and 7. Moreover, in Table 3, the temperatures corresponding to 5, 10, and 50% mass loss and the temperatures at a maximum of mass-loss rate for the EVA-based nanocomposites have been tabulated. It is well-known that the thermal stability of polymer composites plays a crucial role in determining their processing and applications since it affects the final properties of the materials such as the upper limit usage temperature and dimensional stability [59]. It is now well accepted that the improvement in thermal stability of polymer nanocomposites containing CNTs is due to the following factors: barrier effect, the thermal conductivity of CNTs, physical or chemical adsorption, radical scavenging action, and polymer-nanotube interaction. For each polymer-CNT composite, thermal stability may be due to one mechanism or the combined action of several processes, which depend on the different components, microstructures, and exterior conditions [59]. Similarly, also HNTs are found to be a promising additive for improving the thermal stability and flame retardancy of polyolefins and other polymers [19]. The improvement of thermal stability and flame retardancy resulted from the barrier properties of HNTs combined with an encapsulation process of the polymer's degradation products inside the HNT lumens [60]. What is also important, because of their structure and chemical character, HNTs can be more easily dispersed into polyolefin matrices in comparison with other nanofillers, such as montmorillonite (MMT) [16,19,61]. This is a crucial factor for obtaining nanocomposites with better mechanical properties, higher thermal stability, and reduced flammability. Of course, there are studies on nanocomposites based on EVA with the addition of MWCNT [62] to improve the thermal stability, however, so far nobody has studied how a hybrid system of carbon nanotubes and mineral nanotubes can improve both thermal and thermo-oxidative stability. Herein we can see that in general, in the oxidizing atmosphere (Figure 6), a three-step degradation process is observed for all samples, while in an inert atmosphere a two-step degradation process is observed (Figure 7). The TGA thermograms of the EVA show that the first degradation process starts at 250 °C (T_{onset}) and is completed at 390 °C [63]. This process corresponds to the loss of acetic acid [64]. The second step corresponds to the degradation of the polyethylene chains and starts at approximately 410 °C and ends at 465 °C [65]. While the third stage starts (observed only in the oxidizing atmosphere) at approximately 480 °C and ends at 570 °C, in the case of neat EVA, whereas in the case of all composites, it starts at approximately 580 °C and ends at 635 °C. In the case of EVA/HNTs nanocomposites and EVA/CNTs+HNTs hybrid nanocomposites, this third step corresponds to the decomposition of carbonaceous-silicate char [66,67]. While in the case of EVA/CNT nanocomposites this third step corresponds to the decomposition of residue formed in the second step of decomposition. Additionally, in an oxidizing atmosphere, the first two characteristic temperatures ($T_{5\%}$ and $T_{10\%}$) were slightly reduced by the incorporation of HNTs, CNTs, and the mixture of both, while at the $T_{50\%}$ (i.e., at higher temperatures) one can observe the enhancement of thermo-oxidative stability by the incorporation of nanofillers, of even 20 °C for EVA/4 wt.% CNT + HNT. In the case of the analysis conducted in the inert atmosphere, almost all samples exhibited improvement in thermal stability. Only in the case of nanocomposites containing 2 and 4 wt.% of HNTs no improvement was observed. In general, the studies conducted on polymer nanocomposites containing HNTs [27,68,69] explained such deterioration of thermal stability by the physical adsorption of water on the external surface of HNTs, contributing to the degradation of polymer matrix [68]. In turn, Bidsorkhi et al. [40] demonstrated that the improvement in thermal stability is attributed to the homogeneous dispersion of HNTs, which originated from strong hydrogen bonding between surface functional groups of HNTs and vinyl acetate groups of EVA. In the case of obtained EVA/HNT nanocomposites, the interaction between HNTs and EVA were confirmed by FTIR spectroscopy. Figure 8 shows FTIR spectra of the used neat EVA copolymer for nanocomposite preparation. The characteristic absorption peaks at 1238 cm^{-1} and 1735 cm^{-1} are assigned to the C–O and C=O stretching in vinyl acetate, whereas reflections at 2850 cm^{-1} and 2918 cm^{-1} are attributed to the C-H stretching vibration in ethylene. For EVA nanocomposites with the content of 6 and 8 wt.% of HNTs, the peak corresponding to stretching vibrations of C=O of EVA shifts to lower wavenumber, from 1735 to 1726 and 1723 cm^{-1}, respectively. This shift may be attributed to hydrogen bonding interactions between the carbonyl groups (C=O) of EVA and the hydroxyl groups of HNTs.

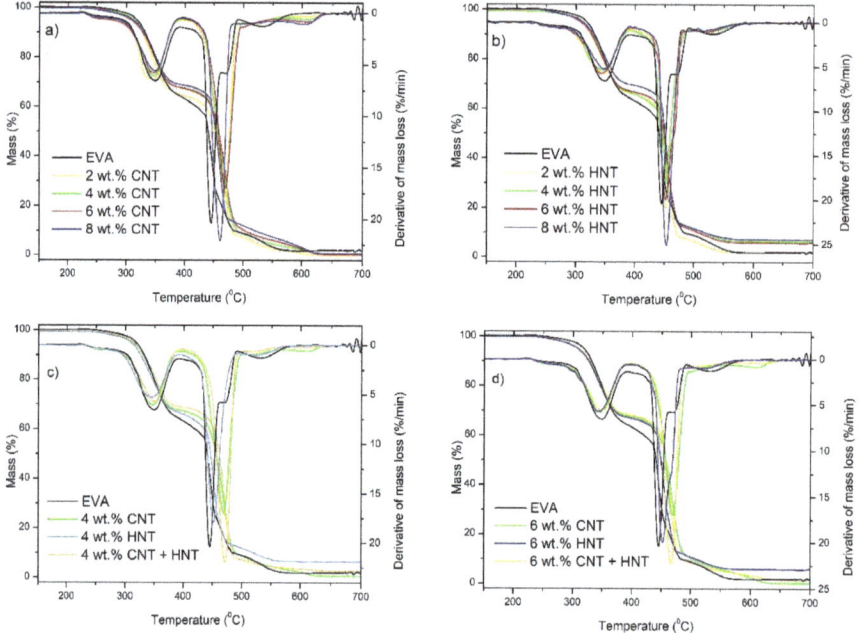

Figure 6. TG and DTG curves of: EVA/MWCNT nanocomposites (**a**); EVA/HNT nanocomposites (**b**); EVA-based nanocomposites at the total nanofillers content of 4 wt.% (**c**); and EVA-based nanocomposites at the total nanofillers content of 6 wt.% (**d**) measured in an oxidizing atmosphere.

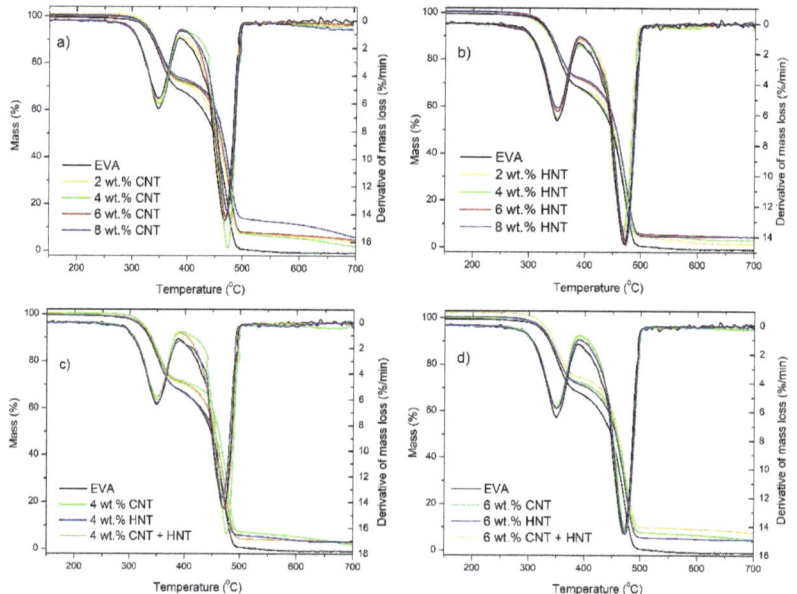

Figure 7. TG and DTG curves of: EVA/MWCNTs nanocomposites (**a**); EVA/HNTs nanocomposites (**b**); EVA-based nanocomposites at the total nanofillers content of 4 wt.% (**c**); and EVA-based nanocomposites at the total nanofillers content of 6 wt.% (**d**) measured in an inert atmosphere.

Table 3. Temperatures corresponding to 5, 10, and 50% mass loss and the temperatures at a maximum of mass loss rate for the EVA-based nanocomposites obtained in an oxidizing and inert atmosphere.

Sample	Air					
	$T_{5\%}$, °C	$T_{10\%}$, °C	$T_{50\%}$, °C	T_{DTG1}, °C	T_{DTG2}, °C	T_{DTG3}, °C
EVA	320	331	441	351	450	535
EVA/2 wt.% CNT	314	328	453	346	488	609
EVA/4 wt.% CNT	314	330	457	348	486	609
EVA/6 wt.% CNT	309	328	459	346	485	609
EVA/8 wt.% CNT	311	330	454	348	483	602
EVA/2wt.% HNT	313	327	449	350	464	526
EVA/4wt.% HNT	310	325	447	346	461	526
EVA/6 wt.% HNT	308	327	450	344	464	525
EVA/8wt.% HNT	311	330	451	350	463	526
EVA/4 wt.% CNT + HNT	315	332	462	351	490	526
EVA/6 wt.% CNT + HNT	310	329	459	351	488	615

Sample	Argon					
	$T_{5\%}$, °C	$T_{10\%}$, °C	$T_{50\%}$, °C	T_{DTG1}, °C	T_{DTG2}, °C	-
EVA	322	336	448	349	470	-
EVA/2 wt.% CNT	330	341	456	350	473	-
EVA/4 wt.% CNT	327	340	462	352	477	-
EVA/6wt.% CNT	327	340	458	349	474	-
EVA/8 wt.% CNT	327	341	463	350	476	-
EVA/2 wt.% HNT	320	334	452	351	471	-
EVA/4 wt.% HNT	320	335	450	350	468	-
EVA/6 wt.% HNT	326	340	456	350	474	-
EVA/8wt.% HNT	325	339	456	351	473	-
EVA/4 wt.%CNT + HNT	324	338	455	351	473	-
EVA/6 wt.% CNT + HNT	333	344	461	351	475	-

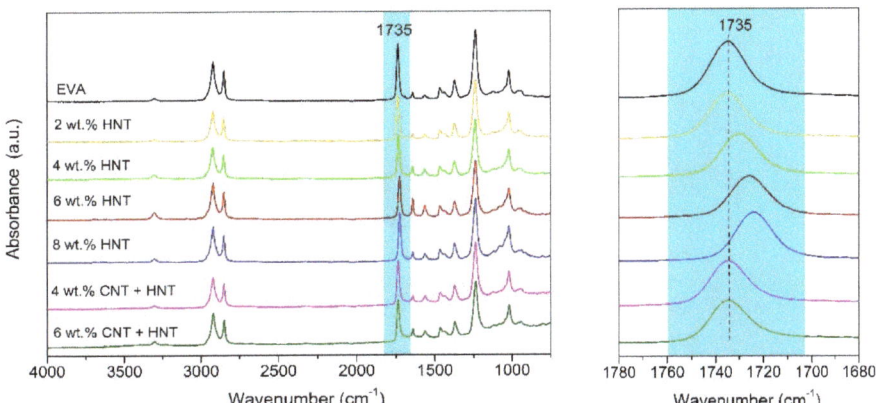

Figure 8. FTIR spectra for the neat EVA copolymer and EVA-based nanocomposites with HNTs and hybrid CNT + HNT.

As a result of the uniform dispersion of HNTs, the highest possible value of the surface-to-volume ratio for the nanofillers is achieved. Therefore, the degraded and/or degrading products of EVA polymer were entrapped within the tubular rods of HNTs, consequently contributing to a delay in the polymer decomposition process. Another possible interpretation for the improvement in the thermal stability of the nanocomposites could be the insulation effect of HNTs. Generally, layered silicates are thought to be an excellent thermal barrier that can effectively protect the matrix from being exposed to heat flow and thermal energy [70]. Although the improvement in thermo-oxidative and thermal stability of polymer/carbon nanotubes and polymer/layered inorganics-based nanocomposites has been reported extensively, the mechanism of such effect is still not yet well understood. Especially when two types of nanofillers, that differ in the properties and structure, are mixed. Herein, both hybrids exhibited a very decent improvement in thermal stability, even though no synergistic effect of property improvement was observed in this case. Generally, the most common explanation suggests that the enhancement in thermal stability derived from the mass and heat transfer barrier caused by a carbonaceous char (CNT-based composites) and carbonaceous-silicate char (HNTs-based composites) on the surface of the polymer melt [59,66,67]. Moreover, in the case of HNTs-nanocomposites, recent studies also suggest that the effect may be associated with a chemical interaction between the polymer matrix and the outer layer surface during thermal degradation and combustion processes [71].

3.4. Physical Properties of EVA-Based Nanocomposites

Physical properties of EVA-based nanocomposites such as density, hardness, melt flow rate, and some mechanical properties are submitted in Table 4 and Figures 9–11. The theoretical and real densities of the nanocomposites were found to increase with nanofillers content because of the higher density of nanotubes over the neat EVA. One can observe that slight deviations of the measured values from the theoretical densities, especially when HNTs were used as filler, were visible. However, the differences are relatively small, and the ratio of experimental density to theoretical density does not exceed 98%. Therefore, it can be concluded that the desired compositions were obtained.

Table 4. Density, hardness, and melt flow rate of EVA/MWCNTs + HNTs nanocomposites.

Material	d_t (g/cm^3)	d_R (g/cm^3)	H (ShD)	MFR (g/10 min)
EVA	0.967	0.973 ± 0.001	22.4 ± 1.7	2.99 ± 0.46
EVA/2 wt.% CNT	0.983	0.980 ± 0.001	26.3 ± 1.3	0.99 ± 0.13
EVA/4 wt.% CNT	0.991	0.989 ± 0.001	27.9 ± 1.1	0.22 ± 0.08
EVA/6 wt.% CNT	1.000	0.999 ± 0.001	30.4 ± 1.0	0.12 ± 0.05
EVA/8 wt.% CNT	1.010	1.009 ± 0.002	33.2 ± 1.4	-
EVA/2 wt.% HNT	0.986	0.982 ± 0.001	22.8 ± 1.0	2.74 ± 0.32
EVA/4 wt.% HNT	0.998	0.989 ± 0.004	23 ± 1.2	4.05 ± 0.27
EVA/6 wt.% HNT	1.011	0.997 ± 0.001	23.1 ± 1.0	3.76 ± 0.46
EVA/8 wt.% HNT	1.024	1.008 ± 0.001	24.5 ± 1.0	3.90 ± 0.18
EVA/4 wt.% CNT + HNT	0.995	0.991 ± 0.001	25.6 ± 0.7	1.87 ± 0.14
EVA/6 wt.% CNT + HNT	1.006	1.001 ± 0.001	25.6 ± 1.5	1.22 ± 0.23

d_t—theoretical density; d_R—density real; H—Shore hardness, scale D; MFR—melt flow rate at 190 °C and 2.16 kg.

The hardness test results show an increase in the Shore D hardness values with increasing fillers content (Table 4). Especially the presence of rigid reinforcement MWCNTs cause a considerable increase in nanocomposites hardness, which for EVA/CNT nanocomposite at a MWCNTs loading of 8 wt.% is approximately 48% higher than those of the neat EVA copolymer, while the hardness of the nanocomposite with 8 wt.% of HNTs is about 9% higher. In turn, as expected, the hybrid nanocomposites have intermediate hardness values fitting between nanocomposites filled with MWCNTs and HNTs.

Since MWCNTs are characterized by high Young's modulus (~1 TPa) and aspect ratio their introduction in the EVA matrix increases the composite stiffness, which can be observed in Figure 10a, as a steepening of stress–strain curves. On the other hand, the filling of EVA with HNT's reduces the stiffness (Figure 10b). The tensile strength of EVA composites increases with the higher filling ratio of MWCNTs, but in case of filling with HNTs the strength is diminished and the increase of filling ratio has little influence on the tensile performance of the composite (Figure 11a). The addition of MWCNTs reinforces the EVA matrix, while HNT filler gives an adverse effect in terms of composite stiffness and tensile strength. However, the HNTs increase to a great extent the strain to break of EVA. The increase of strain at break and a slight decrease of tensile strength is also noticeable at 2 wt.% of MWCNTs filling ratio, but the effect is explained by high stiffness of MWCNTs, therefore it diminishes with higher filling ratios of MWCNTs. Taking into account that HNTs are considered to be fibrous nanoparticles, however with lower aspect ratio and lower stiffness than their carbon counterpart, they are still much stiffer than the neat EVA copolymer. Therefore, the increase of strain to break and reduction of strength and stiffness by filling with HNTs may be an effect of HNTs affecting the supermolecular structure of EVA. This is also relevant to MWCNTs although this effect is covered up by greater gains in terms of reinforcement at filling ratio > 2 wt.%.

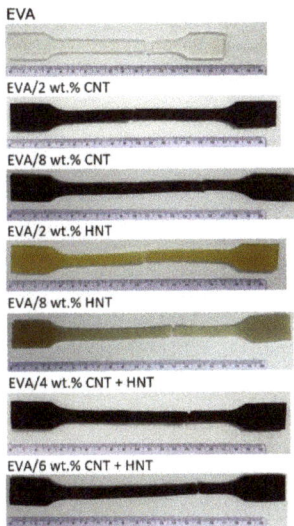

Figure 9. Samples of EVA-based nanocomposites after stretching.

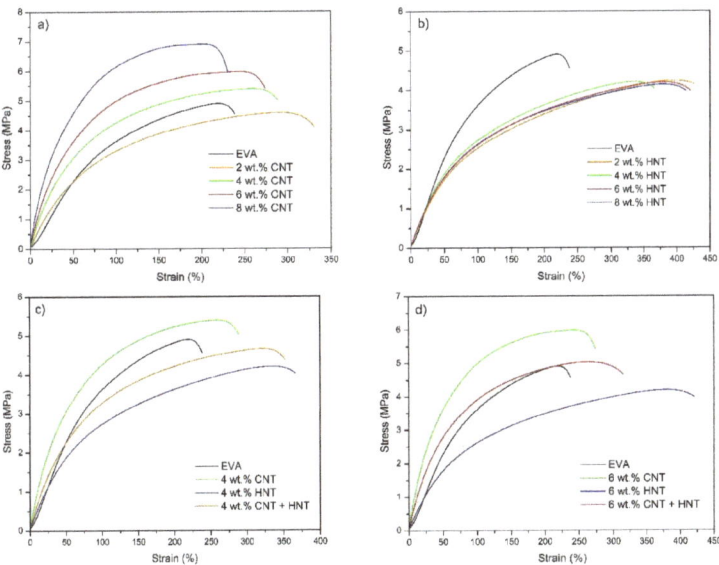

Figure 10. Stress–strain curves of EVA/MWCNT nanocomposites (**a**), EVA/HNT nanocomposites (**b**), EVA-based nanocomposites at the total nanofillers content of 4 wt.% (**c**), and EVA-based nanocomposites at the total nanofillers content of 6 wt.% (**d**).

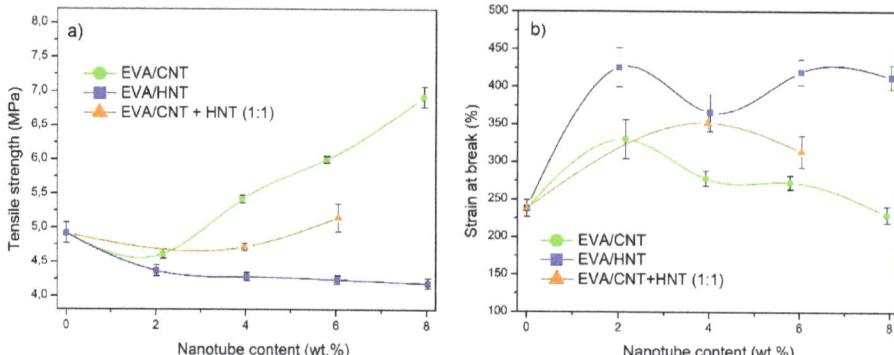

Figure 11. Tensile strength (**a**) and strain at break (**b**) of EVA/MWCNTs + HNTs nanocomposites with respect to nanofillers content.

The polymer crystal phase transformation or polymer chain intercalation leads usually to such effect and has been recorded for some polymer/nanoparticle systems if the good distribution of nanoparticles in the polymer chains is provided, which can be obtained only by good compatibility between the matrix and particles [72,73]. The increase of strain at break was usually obtained by incorporation of layered silicates in such polymers as polyvinylidene fluoride (PVDF), ethylene propylene diene rubber (EPDM), and polyurethanes (PU) [72,74–76]. Among them also PU nanocomposites with HNTs gave an extraordinary coupled increase of strength and strain to break [33]. Also, layered graphene introduced in-situ in polyester altered the same way the stress-strain behavior [77]. This plasticizing effect is also apparent by 15% lower injection molding pressures for the EVA/HNT compounds in comparison to native EVA (Table 1).

The extent degradation of the EVA matrix upon processing may be ruled out as this would also decrease strains to break of manufactured composites. The addition of HNT's increases also the flow of EVA, which was reflected indirectly in the results MFR presented in Table 4 (MFR). However, it cannot be assessed whether this was caused by EVA/HNTs interaction, degradation of EVA, or a combination of both factors. The interaction of HNTs and MWCNTs particles with EVA polymer chains has not been observed in the results of DSC measurements (Table 2), where only minute changes in the enthalpies of crystal melting and their characteristic temperatures were recorded. Nevertheless, in the EVA nanocomposites, where significant alternation of the supermolecular structure was proven, also, changes have not been observed in the DSC measurements [26,65]. It is also important to notice that MWCNTs and HNTs limited the recoiling of the polymer chains as the samples were not returning to their initial dimensions as quickly as for the neat EVA (Figure 9).

To investigate the elastic deformability and reversibility of the obtained nanocomposites, cyclic tensile tests were carried out. Results are presented in Figures 12–14. The contours made by the loops are consistent with the characteristics obtained under the static tensile tests. The hybrid nanocomposites are the exception since tensile strength in the case of cyclic tests is greater than the tensile strength of corresponding nanocomposites with one type of filler (Figure 12). As shown in Figure 13, the value of modulus at 200% strain during cyclic tests, is the highest for hybrid nanocomposite containing 6 wt.% of CNTs and HNTs (over 5.5 MPa). EVA/4 wt.% CNTs + HNTs also achieves higher modulus at 200% strain value than the corresponding nanocomposites with CNTs or HNTs. These differences may result from the orientation of nanotubes along with the tensile direction. After the removal of the applied force, nanotubes did not return to the original position. Thus, the addition of HNTs enables the material to reach higher strains by its plasticizing effect, while CNTs align upon cycling straining and give a higher gain in terms of tensile strength. This reveals an intriguing strengthening mechanism, which occurs because of synergism of both nanotube fillers and which can be used in the development of nanocomposites with behavior allowing to maintain higher cyclic strains and loads.

It can be seen that nanocomposites containing CNTs show higher values of the permanent set than the neat EVA copolymer (Figure 14). The EVA/8wt.% CNT nanocomposite showed the highest value of permanent set (PS(200%) of over 85%, which was about two times higher than that of neat EVA copolymer. As mentioned earlier, CNTs increase the stiffness of the composite, which can contribute to higher residual strain values. HNTs behave the opposite way. As can be seen in Figure 12b, EVA/HNTs nanocomposites have slightly better recovery properties, than the neat EVA copolymer.

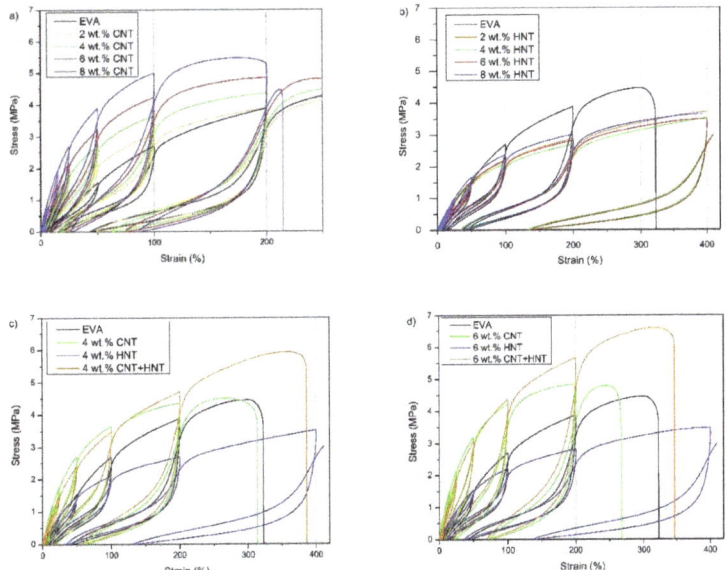

Figure 12. The stress–strain curves under cyclic loading for EVA/MWCNT (**a**), EVA/HNT (**b**), EVA/CNT + HNT (**c**,**d**) nanocomposites, and the neat EVA copolymer under. Results corresponding to 5%, 15%, 25%, 50%, 100%, and 200% maximum tensile strains.

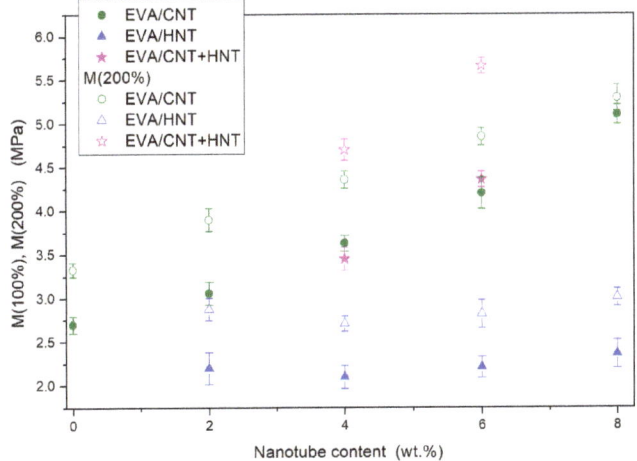

Figure 13. Modulus at 100 and 200% of elongation for the neat EVA and EVA-based nanocomposites after maximum strain attained in the cycle.

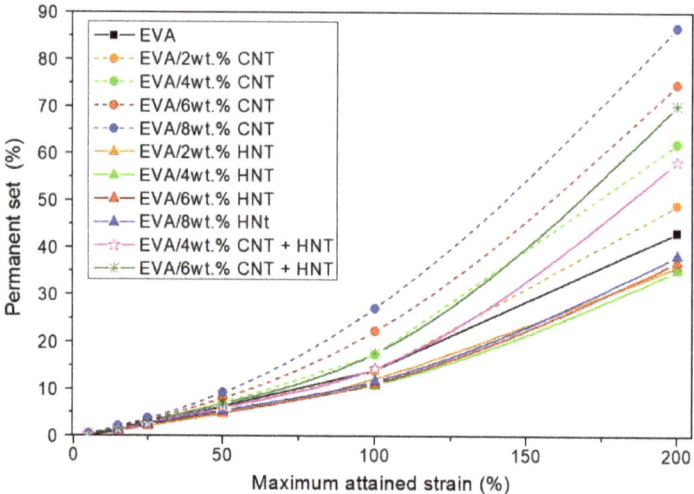

Figure 14. Permanent set values of the neat EVA, and EVA-based nanocomposites after maximum strain attained in a cycle.

3.5. Electrical Conductivity of EVA Nanocomposites

MWCNTs possess very high intrinsic electrical conductivity. Even their small content can significantly improve the electrical conductivity of insulating polymers. Figure 15 displays the electrical properties of EVA-based nanocomposites. It can be seen that the electrical conductivity increased steadily in EVA/CNTs nanocomposites, as the CNTs concentrations increased. The highest electrical conductivity value of 5.2×10^{-7} S/m was achieved for nanocomposite with 8 wt.% of CNTs. It is almost 6 orders of magnitude higher than for neat EVA copolymer. Obtained materials did not show a sharp increase in electrical conductivity along with the increase in MWCNTs content, in contrast to the results from previous work in which the matrix was low-density polyethylene (LDPE) [78]. In LDPE/MWCNTs nanocomposites the percolation threshold was observed at the loading of 1.5 wt.% of the same MWCNTs [78]. It is known that relatively uniform dispersion of CNTs can be achieved in polar polymers such as polyamide, polycarbonate because of the strong interaction between the polar moiety of the polymer chains and the surface of the CNTs [79]. EVA copolymer is polar in comparison to the LDPE and its polarity is dependent on the mass content of VA in the copolymer. Because of the interactions between polar groups of EVA copolymer and nanotubes, the carbon nanotubes may be more concentrated in VA polar domains. Therefore, in EVA nanocomposites with the highest concentration (8 wt.%) of nanotubes, regions/agglomerates with a higher content of entangled of nanotubes are visible (Figure 2).

Because halloysite nanotubes are not conductive fillers, the EVA/HNTs nanocomposites exhibit lower electrical conductivity than a neat EVA copolymer. As shown in Figure 15, the conductivity decreased by two orders of magnitude. Moreover, the insulating properties of HNTs cause no significant improvement in the conductivity of hybrid nanocomposites. Hybrid material at the concentration of 4 wt.% exhibits an even slightly lower value of conductivity if compared to neat EVA. This can be explained by the fact that HNTs located between CNTs, impedes the formation of conductive pathways.

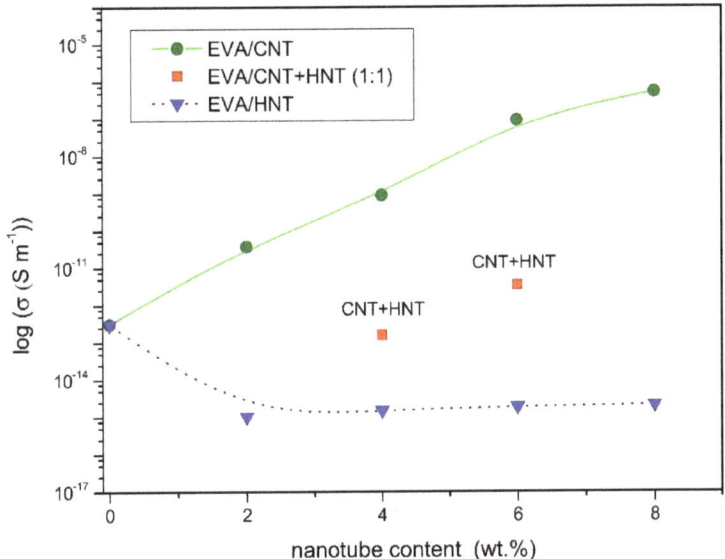

Figure 15. Electrical conductivity of EVA/CNT, EVA/HNT, and EVA/CNT + HNT nanocomposites.

4. Conclusions

Nanocomposites based on EVA copolymer containing MWCNTs, HNTs, or both types of nanotubes were prepared by melt blending. One found that relatively good nanotubes distribution in polymer matrix were obtained. In the case of EVA/CNT nanocomposites at high load (6–8 wt.%), the MWCNTs form the highly exfoliated network structure in EVA matrix, in which individual nanotubes and entangled CNTs in agglomerates are present. DSC studies showed that the addition of nanofillers caused no significant effect on the melting temperature and the degree of crystallinity. However, glass transition temperature slightly increased. At high temperatures ($T_{50\%}$), nanocomposites showed better thermo-oxidative stability than the neat EVA copolymer. A slight improvement in thermal stability was also noted. Moreover, the addition of CNTs has significantly improved the mechanical properties of EVA copolymer. Nanocomposites were stiffer and their tensile strength increased by about 40%. In turn, HNTs give an opposite effect in terms of composite stiffness and tensile strength. They decrease the strength of EVA. However, the strain to break increases by over 70% when HNTs are added. Furthermore, the cyclic tensile tests demonstrated that nanocomposites with HNTs have slightly better recovery properties, than pure EVA. Interestingly, in cyclic tensile tests significant improvement of tensile strength for hybrid nanocomposites was visible. Moreover, the extraordinary strengthening caused by synergism of the used nanotube fillers did not diminish the strain rates achieved by the hybrids. Nanocomposites with CNTs were found to be electrically conducting. For nanocomposites containing 8 wt.% of CNTs, an increase in electrical conductivity for about six orders of magnitude in comparison to the neat copolymer was observed.

Supplementary Materials: The following are available online at http://www.mdpi.com/1996-1944/13/17/3809/s1. Figure S1: TEM images of EVA/6 wt.% CNT + HNT nanocomposite at different magnifications. Figure S2: TEM images of EVA/6 wt.% CNTs nanocomposite at different magnifications.

Author Contributions: A.Z. prepared the literature review, analyzed the results, wrote the original draft of the manuscript, and performed physical properties measurements; A.S. planned the experiment, supervised the discussion, and reviewed the manuscript; P.F. prepared the samples, analyzed the mechanical results, and reviewed the manuscript; A.K. performed the SEM experiments; I.J. participated in TEM analysis and supervised the discussion of the results; S.P. analyzed the thermal properties of the samples and participated in editing and revision of the manuscript. All authors have read and agreed to the published version of the manuscript.

Funding: This research received no external funding.

Acknowledgments: The authors wish to thank Walid Baazis from IPCMS (UMR 7504 CNRS-UDS, Strasbourg) for the TEM investigations. The authors also would like to express their appreciation to PRACHT GROUP for the long-term loan of Arburg Allrounder 270 S 350–100 injection molding machine.

Conflicts of Interest: The authors declare no conflict of interest.

References

1. Iijima, S. Helical microtubules of graphitic carbon. *Nature* **1991**, *354*, 56–58. [CrossRef]
2. Saifuddin, N.; Raziah, A.Z.; Junizah, A.R. Carbon Nanotubes: A Review on Structure and Their Interaction with Proteins. *J. Chem.* **2013**, *2013*, 676815. [CrossRef]
3. Li, Y.; Huang, X.; Zeng, L.; Li, R.; Tian, H.; Fu, X.; Wang, Y.; Zhong, W.-H. A review of the electrical and mechanical properties of carbon nanofiller-reinforced polymer composites. *J. Mater. Sci.* **2019**, *54*, 1036–1076. [CrossRef]
4. Ates, M.; Eker, A.A.; Eker, B. Carbon nanotube-based nanocomposites and their applications. *J. Adhes. Sci. Technol.* **2017**, *31*, 1977–1997. [CrossRef]
5. Mittal, G.; Dhand, V.; Rhee, K.Y.; Park, S.-J.; Lee, W.R. A review on carbon nanotubes and graphene as fillers in reinforced polymer nanocomposites. *J. Ind. Eng. Chem.* **2015**, *21*, 11–25. [CrossRef]
6. Ma, P.C.; Kim, J.-K.; Tang, B.Z. Effects of silane functionalization on the properties of carbon nanotube/epoxy nanocomposites. *Compos. Sci. Technol.* **2007**, *67*, 2965–2972. [CrossRef]
7. Song, Y.S.; Youn, J.R. Influence of dispersion states of carbon nanotubes on physical properties of epoxy nanocomposites. *Carbon N.Y.* **2005**, *43*, 1378–1385. [CrossRef]
8. Ayatollahi, M.R.; Shadlou, S.; Shokrieh, M.M.; Chitsazzadeh, M. Effect of multi-walled carbon nanotube aspect ratio on mechanical and electrical properties of epoxy-based nanocomposites. *Polym. Test.* **2011**, *30*, 548–556. [CrossRef]
9. Gorrasi, G.; Sarno, M.; Di Bartolomeo, A.; Sannino, D.; Ciambelli, P.; Vittoria, V. Incorporation of carbon nanotubes into polyethylene by high energy ball milling: Morphology and physical properties. *J. Polym. Sci. Part B Polym. Phys.* **2007**, *45*, 597–606. [CrossRef]
10. Mierczynska, A.; Mayne-L'Hermite, M.; Boiteux, G.; Jeszka, J.K. Electrical and mechanical properties of carbon nanotube/ultrahigh-molecular-weight polyethylene composites prepared by a filler prelocalization method. *J. Appl. Polym. Sci.* **2007**, *105*, 158–168. [CrossRef]
11. Ngabonziza, Y.; Li, J.; Barry, C.F. Electrical conductivity and mechanical properties of multiwalled carbon nanotube-reinforced polypropylene nanocomposites. *Acta. Mech.* **2011**, *220*, 289–298. [CrossRef]
12. Wu, H.-Y.; Jia, L.-C.; Yan, D.-X.; Gao, J.; Zhang, X.-P.; Ren, P.-G.; Li, Z.-M. Simultaneously improved electromagnetic interference shielding and mechanical performance of segregated carbon nanotube/polypropylene composite via solid phase molding. *Compos. Sci. Technol.* **2018**, *156*, 87–94. [CrossRef]
13. Verma, P.; Saini, P.; Choudhary, V. Designing of carbon nanotube/polymer composites using melt recirculation approach: Effect of aspect ratio on mechanical, electrical and EMI shielding response. *Mater. Des.* **2015**, *88*, 269–277. [CrossRef]
14. Jiang, F.; Zhang, L.; Jiang, Y.; Lu, Y.; Wang, W. Effect of annealing treatment on the structure and properties of polyurethane/multiwalled carbon nanotube nanocomposites. *J. Appl. Polym. Sci.* **2012**, *126*, 845–852. [CrossRef]
15. Kwon, J.; Kim, H. Comparison of the properties of waterborne polyurethane/multiwalled carbon nanotube and acid-treated multiwalled carbon nanotube composites prepared by in situ polymerization. *J. Polym. Sci. Part A Polym. Chem.* **2005**, *43*, 3973–3985. [CrossRef]
16. Liu, M.; Jia, Z.; Jia, D.; Zhou, C. Recent advance in research on halloysite nanotubes-polymer nanocomposite. *Prog. Polym. Sci.* **2014**, *39*, 1498–1525. [CrossRef]
17. Bordeepong, S.; Bhongsuwan, D.; Pungrassami, T.; Bhongsuwan, T. Characterization of halloysite from Thung Yai District, Nakhon Si Thammarat Province, in Southern Thailand. *Songklanakarin J. Sci. Technol.* **2011**, *33*, 599–607.

18. Padhi, S.; Ganga, P.; Achary, R.; Nayak, N.C. Mechanical and morphological properties of halloysite nanotubes filled ethylene-vinyl acetate copolymer nanocomposites. *Indian J. Chem. Technol.* **2017**, *24*, 184–191. [CrossRef]
19. Szpilska, K.; Czaja, K.; Kudła, S. Halloysite nanotubes as polyolefin fillers. *Polim. Polym.* **2015**, *60*, 359–371. [CrossRef]
20. Huang, J.; Tang, Z.; Guo, B. Surface Modification of Halloysite. In *Functional Polymer Composites with Nanoclays*; Lvov, Y., Guo, B., Fakhrullin, R.F., Eds.; Royal Society of Chemistry: London, UK, 2016; pp. 157–186. ISBN 978-1-78262-672-5.
21. Liu, M.; Guo, B.; Du, M.; Chen, F.; Jia, D. Halloysite Nanotubes as a Novel β-Nucleating Agent for Isotactic Polypropylene. *Polymer* **2009**, *50*, 3022–3030. [CrossRef]
22. Prashantha, K.; Schmitt, H.; Lacrampe, M.F.; Krawczak, P. Mechanical behaviour and essential work of fracture of halloysite nanotubes filled polyamide 6 nanocomposites. *Compos. Sci. Technol.* **2011**, *71*, 1859–1866. [CrossRef]
23. Prashantha, K.; Lacrampe, M.F.; Krawczak, P. Processing and characterization of halloysite nanotubes filled polypropylene nanocomposites based on a masterbatch route: Effect of halloysites treatment on structural and mechanical properties. *Express Polym. Lett.* **2011**, *5*, 295–307. [CrossRef]
24. Pedrazzoli, D.; Pegoretti, A.; Thomann, R.; Kristóf, J.; Karger-Kocsis, J. Toughening linear low-density polyethylene with halloysite nanotubes. *Polym. Compos.* **2015**, *36*, 869–883. [CrossRef]
25. Singh, V.P.; Vimal, K.K.; Kapur, G.S.; Sharma, S.; Choudhary, V. High-density polyethylene/halloysite nanocomposites: Morphology and rheological behaviour under extensional and shear flow. *J. Polym. Res.* **2016**, *23*, 43. [CrossRef]
26. Ning, N.; Yin, Q.; Luo, F.; Zhang, Q.; Du, R.; Fu, Q. Crystallization behavior and mechanical properties of polypropylene/halloysite composites. *Polymer* **2007**, *48*, 7374–7384. [CrossRef]
27. Wang, B.; Huang, H.-X. Effects of halloysite nanotube orientation on crystallization and thermal stability of polypropylene nanocomposites. *Polym. Degrad. Stab.* **2013**, *98*, 1601–1608. [CrossRef]
28. Liu, M.; Guo, B.; Du, M.; Lei, Y.; Jia, D. Natural Inorganic Nanotubes Reinforced Epoxy Resin Nanocomposites. *J. Polym. Res.* **2007**, *15*, 205–212. [CrossRef]
29. Deng, S.; Zhang, J.; Ye, L.; Wu, J. Toughening epoxies with halloysite nanotubes. *Polymer* **2008**, *49*, 5119–5127. [CrossRef]
30. Jen, Y.-M.; Huang, J.-C. Synergistic Effect on the Thermomechanical and Electrical Properties of Epoxy Composites with the Enhancement of Carbon Nanotubes and Graphene Nano Platelets. *Materials* **2019**, *12*, 255. [CrossRef]
31. Silva, M.; Vale, D.; Rocha, J.; Rocha, N.; Santos, R.M. Synergetic effects of carbon nanotube-graphene nanoplatelet hybrids in carbon fibre reinforced polymer composites. *MATEC Web Conf.* **2018**, *188*, 1015. [CrossRef]
32. Paszkiewicz, S.; Szymczyk, A.; Sui, X.M.; Wagner, H.D.; Linares, A.; Ezquerra, T.A.; Rosłaniec, Z. Synergetic effect of single-walled carbon nanotubes (SWCNT) and graphene nanoplatelets (GNP) in electrically conductive PTT-block-PTMO hybrid nanocomposites prepared by in situ polymerization. *Compos. Sci. Technol.* **2015**, *118*, 72–77. [CrossRef]
33. Jiang, L.; Zhang, C.; Liu, M.; Yang, Z.; Tjiu, W.W.; Liu, T. Simultaneous reinforcement and toughening of polyurethane composites with carbon nanotube/halloysite nanotube hybrids. *Compos. Sci. Technol.* **2014**, *91*, 98–103. [CrossRef]
34. Shahneel Saharudin, M.; Atif, R.; Hasbi, S.; Naguib Ahmad Nazri, M.; Ubaidah Saidin, N.; Abdullah, Y. Synergistic effects of halloysite and carbon nanotubes (HNTs + CNTs) on the mechanical properties of epoxy nanocomposites. *AIMS Mater. Sci.* **2019**, *6*, 900–910. [CrossRef]
35. Franciszczak, P.; Bledzki, A.K. Tailoring of dual-interface in high tenacity PP composites—Toughening with positive hybrid effect. *Compos. Part. A Appl. Sci. Manuf.* **2016**, *83*, 185–192. [CrossRef]
36. Franciszczak, P.; Merijs-Meri, R.; Kalniņš, K.; Błędzki, A.K.; Zicans, J. Short-fibre hybrid polypropylene composites reinforced with PET and Rayon fibres–Effects of SSP and interphase tailoring. *Compos. Struct.* **2017**, *181*, 121–137. [CrossRef]
37. Franciszczak, P.; Kalniņš, K.; Błędzki, A.K. Hybridisation of man-made cellulose and glass reinforcement in short-fibre composites for injection moulding–Effects on mechanical performance. *Compos. Part. B Eng.* **2018**, *145*, 14–27. [CrossRef]

38. Sahu, S.K.; Badgayan, N.D.; Samanta, S.; Rama Sreekanth, P.S. Quasistatic and dynamic nanomechanical properties of HDPE reinforced with 0/1/2 dimensional carbon nanofillers based hybrid nanocomposite using nanoindentation. *Mater. Chem. Phys.* **2018**, *203*, 173–184. [CrossRef]
39. Tarawneh, M.A.; Chen, R.S.; Hj Ahmad, S.; Al-Tarawni, M.A.M.; Saraireh, S.A. Hybridization of a thermoplastic natural rubber composite with multi-walled carbon nanotubes/silicon carbide nanoparticles and the effects on morphological, thermal, and mechanical properties. *Polym. Compos.* **2019**, *40*, E695–E703. [CrossRef]
40. Bidsorkhi, H.C.; Adelnia, H.; Heidar Pour, R.; Soheilmoghaddam, M. Preparation and characterization of ethylene-vinyl acetate/halloysite nanotube nanocomposites. *J. Mater. Sci.* **2015**, *50*, 3237–3245. [CrossRef]
41. Sabet, M.; Soleimani, H.; Hosseini, S. Properties and characterization of ethylene-vinyl acetate filled with carbon nanotube. *Polym. Bull.* **2016**, *73*, 419–434. [CrossRef]
42. Gaidukovs, S.; Zukulis, E.; Bochkov, I.; Vaivodiss, R.; Gaidukova, G. Enhanced mechanical, conductivity, and dielectric characteristics of ethylene vinyl acetate copolymer composite filled with carbon nanotubes. *J. Thermoplast. Compos. Mater.* **2018**, *31*, 1161–1180. [CrossRef]
43. Park, K.-W.; Kim, G.-H. Ethylene vinyl acetate copolymer (EVA)/multiwalled carbon nanotube (MWCNT) nanocomposite foams. *J. Appl. Polym. Sci.* **2009**, *112*, 1845–1849. [CrossRef]
44. Stanciu, N.V.; Stan, F.; Sandu, I.L.; Susac, F.; Fetecau, C.; Rosculet, R.T. Mechanical, Electrical and Rheological Behavior of Ethylene-Vinyl Acetate/Multi-Walled Carbon Nanotube Composites. *Polymers* **2019**, *11*, 1300. [CrossRef] [PubMed]
45. Kurup, S.N.; Ellingford, C.; Wan, C. Shape memory properties of polyethylene/ethylene vinyl acetate/carbon nanotube composites. *Polym. Test.* **2020**, *81*, 106227. [CrossRef]
46. Wunderlich, B.; Dole, M. Specific heat of synthetic high polymers. VIII. Low pressure polyethylene. *J. Polym. Sci.* **1957**, *24*, 201–213. [CrossRef]
47. Drits, V.A.; Sakharov, B.A.; Hillier, S. Phase and structural features of tubular halloysite (7 Å). *Clay Miner.* **2018**, *53*, 691–720. [CrossRef]
48. Che, B.D.; Nguyen, B.Q.; Nguyen, L.T.T.; Nguyen, H.T.; Nguyen, V.Q.; Van Le, T.; Nguyen, N.H. The impact of different multi-walled carbon nanotubes on the X-band microwave absorption of their epoxy nanocomposites. *Chem. Cent. J.* **2015**, *9*, 10. [CrossRef]
49. Yao, X.; Wu, H.; Wang, J.; Qu, S.; Chen, G. Carbon Nanotube/Poly(methyl methacrylate) (CNT/PMMA) Composite Electrode Fabricated by In Situ Polymerization for Microchip Capillary Electrophoresis. *Chem. A Eur. J.* **2007**, *13*, 846–853. [CrossRef]
50. Wu, J.; Xiang, F.; Han, L.; Huang, T.; Wang, Y.; Peng, Y.; Wu, H. Effects of carbon nanotubes on glass transition and crystallization behaviors in immiscible polystyrene/polypropylene blends. *Polym. Eng. Sci.* **2011**, *51*, 585–591. [CrossRef]
51. Szymczyk, A. Poly(trimethylene terephthalate-block-tetramethylene oxide) elastomer/single-walled carbon nanotubes nanocomposites: Synthesis, structure, and properties. *J. Appl. Polym. Sci.* **2012**, *126*, 796–807. [CrossRef]
52. Li, L.; Li, C.Y.; Ni, C.; Rong, L.; Hsiao, B. Structure and crystallization behavior of Nylon 66/multi-walled carbon nanotube nanocomposites at low carbon nanotube contents. *Polymer* **2007**, *48*, 3452–3460. [CrossRef]
53. Kim, J.Y.; Park, H.S.; Kim, S.H. Unique nucleation of multi-walled carbon nanotube and poly(ethylene 2,6-naphthalate) nanocomposites during non-isothermal crystallization. *Polymer* **2006**, *47*, 1379–1389. [CrossRef]
54. Pérez, R.A.; López, J.V.; Hoskins, J.N.; Zhang, B.; Grayson, S.M.; Casas, M.T.; Puiggalí, J.; Müller, A.J. Nucleation and Antinucleation Effects of Functionalized Carbon Nanotubes on Cyclic and Linear Poly(ε-caprolactones). *Macromolecules* **2014**, *47*, 3553–3566. [CrossRef]
55. Müller, A.J.; Arnal, M.L.; Trujillo, M.; Lorenzo, A.T. Super-nucleation in nanocomposites and confinement effects on the crystallizable components within block copolymers, miktoarm star copolymers and nanocomposites. *Eur. Polym. J.* **2011**, *47*, 614–629. [CrossRef]
56. Goh, H.W.; Goh, S.H.; Xu, G.Q.; Pramoda, K.P.; Zhang, W.D. Crystallization and dynamic mechanical behavior of double-C60-end-capped poly(ethylene oxide)/multi-walled carbon nanotube composites. *Chem. Phys. Lett.* **2003**, *379*, 236–241. [CrossRef]

57. Szymczyk, A.; Roslaniec, Z.; Zenker, M.; Garcia-Gutierrez, M.C.; Hernandez, J.J.; Rueda, D.R.; Nogales, A.; Ezquerra, T.A. Preparation and characterization of nanocomposites based on COOH functionalized multi-walled carbon nanotubes and on poly(trimethylene terephthalate). *Express Polym. Lett.* **2011**, *5*, 977–995. [CrossRef]
58. Paszkiewicz, S.; Szymczyk, A.; Zubkiewicz, A.; Subocz, J.; Stanik, R.; Szczepaniak, J. Enhanced Functional Properties of Low-Density Polyethylene Nanocomposites Containing Hybrid Fillers of Multi-Walled Carbon Nanotubes and Nano Carbon Black. *Polymers* **2020**, *12*, 1356. [CrossRef]
59. Su, S.P.; Xu, Y.H.; Wilkie, C.A. Thermal degradation of polymer-carbon nanotube composites. *Polym. Nanotub. Compos. Prep. Prop. Appl.* **2011**, *2*, 482–510.
60. Rybinski, P.; Grazyna, J. Flammability and other properties of elastomeric materials and nano-materials. Part, I. Nanocomposites of elastomers with montmorillonite or halloysite. *Polim. Polym.* **2013**, *58*, 325–420.
61. Zubkiewicz, A.; Szymczyk, A.; Paszkiewicz, S.; Jędrzejewski, R.; Piesowicz, E.; Siemiński, J. Ethylene vinyl acetate copolymer/halloysite nanotubes nanocomposites with enhanced mechanical and thermal properties. *J. Appl. Polym. Sci.* **2020**, *17*, 49135. [CrossRef]
62. Ye, L.; Wu, Q.; Qu, B. Synergistic effects and mechanism of multiwalled carbon nanotubes with magnesium hydroxide in halogen-free flame retardant EVA/MH/MWNT nanocomposites. *Polym. Degrad. Stab.* **2009**, *94*, 751–756. [CrossRef]
63. Omastová, M.; Číková, E.; Mičušík, M. Electrospinning of ethylene vinyl acetate/carbon nanotube nanocomposite fibers. *Polymers* **2019**, *11*, 550. [CrossRef] [PubMed]
64. Maurin, M.B.; Dittert, L.W.; Hussain, A.A. Thermogravimetric analysis of ethylene-vinyl acetate copolymers with Fourier transform infrared analysis of the pyrolysis products. *Thermochim. Acta* **1991**, *186*, 97–102. [CrossRef]
65. Costache, M.C.; Jiang, D.D.; Wilkie, C.A. Thermal degradation of ethylene-vinyl acetate coplymer nanocomposites. *Polymer* **2005**, *46*, 6947–6958. [CrossRef]
66. Zanetti, M.; Camino, G.; Thomann, R.; Mülhaupt, R. Synthesis and thermal behaviour of layered silicate–EVA nanocomposites. *Polymer* **2001**, *42*, 4501–4507. [CrossRef]
67. Gilman, J.W.; Jackson, C.L.; Morgan, A.B.; Harris, R.; Manias, E.; Giannelis, E.P.; Wuthenow, M.; Hilton, D.; Phillips, S.H. Flammability properties of polymer—Layered-silicate nanocomposites. Polypropylene and polystyrene nanocomposites. *Chem. Mater.* **2000**, *12*, 1866–1873. [CrossRef]
68. Carli, L.N.; Crespo, J.S.; Mauler, R.S. PHBV nanocomposites based on organomodified montmorillonite and halloysite: The effect of clay type on the morphology and thermal and mechanical properties. *Compos. Part. A Appl. Sci. Manuf.* **2011**, *42*, 1601–1608. [CrossRef]
69. Ismail, H.; Pasbakhsh, P.; Fauzi, M.N.A.; Abu Bakar, A. Morphological, thermal and tensile properties of halloysite nanotubes filled ethylene propylene diene monomer (EPDM) nanocomposites. *Polym. Test.* **2008**, *27*, 841–850. [CrossRef]
70. Du, M.; Guo, B.; Jia, D. Thermal stability and flame retardant effects of halloysite nanotubes on poly(propylene). *Eur. Polym. J.* **2006**, *42*, 1362–1369. [CrossRef]
71. Qiu, L.; Chen, W.; Qu, B. Morphology and thermal stabilization mechanism of LLDPE/MMT and LLDPE/LDH nanocomposites. *Polymer* **2006**, *47*, 922–930. [CrossRef]
72. Shah, D.; Maiti, P.; Gunn, E.; Schmidt, D.F.; Jiang, D.D.; Batt, C.A.; Giannelis, E.P. Dramatic Enhancements in Toughness of Polyvinylidene Fluoride Nanocomposites via Nanoclay-Directed Crystal Structure and Morphology. *Adv. Mater.* **2004**, *16*, 1173–1177. [CrossRef]
73. Xie, X.-L.; Li, R.K.-Y.; Liu, Q.-X.; Mai, Y.-W. Structure-property relationships of in-situ PMMA modified nano-sized antimony trioxide filled poly(vinyl chloride) nanocomposites. *Polymer* **2004**, *45*, 2793–2802. [CrossRef]
74. Chang, Y.-W.; Yang, Y.; Ryu, S.; Nah, C. Preparation and properties of EPDM/organomontmorillonite hybrid nanocomposites. *Polym. Int.* **2002**, *51*, 319–324. [CrossRef]
75. Lee, H.-T.; Lin, L.-H. Waterborne Polyurethane/Clay Nanocomposites: Novel Effects of the Clay and Its Interlayer Ions on the Morphology and Physical and Electrical Properties. *Macromolecules* **2006**, *39*, 6133–6141. [CrossRef]

76. Burmistr, M.V.; Sukhyy, K.M.; Shilov, V.V.; Pissis, P.; Spanoudaki, A.; Sukha, I.V.; Tomilo, V.I.; Gomza, Y.P. Synthesis, structure, thermal and mechanical properties of nanocomposites based on linear polymers and layered silicates modified by polymeric quaternary ammonium salts (ionenes). *Polymer* **2005**, *46*, 12226–12232. [CrossRef]
77. Paszkiewicz, S.; Janowska, I.; Pawlikowska, D.; Szymczyk, A.; Irska, I.; Lisiecki, S.; Stanik, R.; Gude, M.; Piesowicz, E. New functional nanocomposites based on poly(trimethylene 2,5-furanoate) and few layer graphene prepared by in situ polymerization. *Express Polym. Lett.* **2018**, *12*, 530–542. [CrossRef]
78. Paszkiewicz, S.; Szymczyk, A.; Pawlikowska, D.; Subocz, J.; Zenker, M.; Masztak, R. Electrically and thermally conductive low density polyethylene-based nanocomposites reinforced by MWCNT or hybrid MWCNT/graphene nanoplatelets with improved thermo-oxidative stability. *Nanomaterials* **2018**, *8*, 264. [CrossRef]
79. Ma, P.-C.; Siddiqui, N.A.; Marom, G.; Kim, J.-K. Dispersion and functionalization of carbon nanotubes for polymer-based nanocomposites: A review. *Compos. Part. A Appl. Sci. Manuf.* **2010**, *41*, 1345–1367. [CrossRef]

© 2020 by the authors. Licensee MDPI, Basel, Switzerland. This article is an open access article distributed under the terms and conditions of the Creative Commons Attribution (CC BY) license (http://creativecommons.org/licenses/by/4.0/).

Article

Impact of the Carbon Nanofillers Addition on Rheology and Absorption Ability of Composite Shear Thickening Fluids

Paulina Nakonieczna-Dąbrowska *, Rafał Wróblewski, Magdalena Płocińska and Marcin Leonowicz

Faculty of Materials Science and Engineering, Warsaw University of Technology, Woloska 141, 02-507 Warsaw, Poland; rafal.wroblewski@pw.edu.pl (R.W.); magdalena.plocinska@pw.edu.pl (M.P.); marcin.leonowicz@pw.edu.pl (M.L.)
* Correspondence: paulina.nakonieczna.dokt@pw.edu.pl

Received: 10 August 2020; Accepted: 27 August 2020; Published: 2 September 2020

Abstract: Synthesis and characterization of composite shear thickening fluids (STFs) containing carbon nanofillers are presented. Shear thickening fluids have attracted particular scientific and technological interest due to their unique ability to abruptly increase viscosity in the case of a sudden impact. The fluids have been developed as a potential component of products with high energy absorbing efficiency. This study reports on the rheological behavior, stability, and microstructure of the STFs modified with the following carbon nanofillers: multi-walled carbon nanotubes, reduced graphene oxide, graphene oxide, and carbon black. In the current experiment, the basic STF was made as a suspension of silica particles with a diameter of 500 nm in polypropylene glycol and with a molar mass of 2000 g/mol. The STF was modified with carbon nanofillers in the following proportions: 0.05, 0.15, and 0.25 vol.%. The addition of the carbon nanofillers modified the rheological behavior and impact absorption ability; for the STF containing 0.25 vol.% of carbon nanotubes, an increase of force absorption up to 12% was observed.

Keywords: shear thickening fluids; nanocomposite fluids; multi-walled carbon nanotubes; carbon fillers

1. Introduction

A fluid is defined as a substance that flows [1]. All fluids that do not comply with this law are called non-Newtonian, i.e., their viscosity depends on the rate and time of shearing. Some paints or blood are examples of such fluids [2]. Non-Newtonian fluids are divided into the following two groups: Rheologically stable which do not change rheological properties during shearing and rheologically unstable with rheological properties that are a function of shear rate and time.

A shear thickening fluid (STF) is composed of the following two basic components: The continuous viscous component which is often various types of glycols and the rigid component which is often silica powder. A suspension of both components brings about a unique effect of shear thickening, resulting in an abrupt growth of viscosity with increasing shear rate [3–5]. Shear thickening fluids display a dilation phenomenon which means that, at a sufficiently high shear rate, their properties change abruptly from that typical of a viscous liquid to the characteristics of an elastic solid [6]. Usually, the STFs exhibit a viscoelastic behavior. The properties of such fluids can be controlled by their composition, which allows them to tailor the final viscosity and critical shear rate at which the initial viscosity abruptly increases [7].

There are several rheological models that, in a more or less precise way, describe the behavior of a particular rheological system in the conditions of low and high shear rates [7]. Each time, an appropriate model is selected to describe a system, it should be accompanied by gathering information on the composition of the fluid and possible interactions among the components of the suspension [8].

The basic mechanisms that describe the phenomena occurring in shear thickening fluids, are primarily based on the frictional forces between the individual solid particles in the suspension [7,9]. According to Reynolds theory, in a basic state, the solid phase particles are softly packed at a low shear rate. The frictional forces between the particles are small, because the liquid between them acts as a lubricant. As the shear rate increases, the distance between the particles decreases (see Figure 1). The slipping ability of the matrix decreases, the friction between the particles increases, as well as the viscosity of the system. Reynolds explained the phenomenon as dilation, based on the change of the volume of the system [7,8]. Another theory that attempts to explain the phenomenon of shear thickening is the order-disorder theory. In agreement with this mechanism, for low shear rates, the solid particles of the STF move in an orderly manner so that the viscosity of the system is low, reflecting the lack of friction between the particles. With an increase of the shear rate, the ordered structure is destroyed, the friction between particles grows, and the viscosity of the system increases. The theory of clusters formation considers mutual hydrodynamic interactions of solid phase particles and collective blocking of their movement as a probable mechanism for the viscosity change [10].

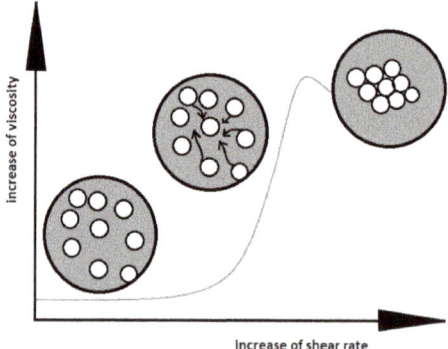

Figure 1. Change of the arrangement of particles in shear thickening fluid with an increase of the shear rate.

The existing mechanisms facilitate, more or less, an understanding of the relationship between the reinforcement (solid phase particles) and the matrix (liquid phase). It is important to tailor the properties of the STF for a particular application. Shear thickening fluids in combination with high-performance woven fabrics, such as para-aramid fabrics (Kevlar and Twaron) or ultra-high molecular weight polyethylene (Dyneema) have been used for a variety of applications [11–13]. The widest application area is related to personal protective products such as helmets, protective sport garments, and smart body armor [14–16].

A variety of modifications have been introduced for the fabrication of composite STFs with different rheological properties. Ge et al. modified STFs with SiC nanowires [17]. Other studies have reported a minor addition of graphene and silica microspheres [18–20], in which the level of maximal viscosities obtained was below 1000 Pa·s. Carbon additives seem to be the best fillers that has improved the maximal viscosity of STFs and also their stability [21–24]. For example, it has been reported that the addition of expanded graphite increased the dilatant effect, up to three times, due to the improved packing of the solid phase [25].

The current study reports on the rheological behavior, stability, and microstructure of the STFs modified with selected carbon nanofillers, such as multi-walled carbon nanotubes, reduced graphene oxide, graphene oxide, and carbon black. The correlations among the properties and structure of the STFs are presented and discussed.

2. Materials and Methods

For the synthesis of shear thickening fluids, an amorphous silica KE-P50, having spherical particles with diameters ranging from 500 to 600 nm, from Nippon Shokubai (Osaka, Japan), were mixed, in appropriate proportions, with polypropylene glycol PPG2000 from Acros Organics (Geel, Belgium), having a molar mass 2000 g/mol and density of 1 ± 0.01 g/cm^3. The content of silica in the basic fluid was 55 vol.% This proportion was chosen on the basis of our previous research, in which silica content ranging from 30 to 60 vol.% were tested (see [26]). A 55 vol.% content of silica obtained the maximal peak of the viscosity and allowed the components to effectively disperse.

The composition of STFs was modified by the addition of the following carbon nanofillers: multi-walled carbon nanotubes (MWCNT, Nanocyl NC7000, Sambreville, Belgium), carbon black (CB, The Cary Company, Addison, IL, USA), graphene oxide (GO, Institute of Electronic Materials Technology, Warsaw, Poland), and reduced graphene oxide (rGO, Institute of Electronic Materials Technology, Warsaw, Poland). The properties of the carbon materials used are shown in Table 1. The STFs were modified by minor additions of the nanofillers with the amounts of 0.05, 0.15 and 0.25 vol%. It was impossible to obtain a homogeneous suspension with the higher content of the nanofillers. In order to disperse the particles in the liquid, the calender EXAKT 80E (Oklahoma City, OK, USA) was used. The synthesis of composite shear thickening fluids consisted of the following two major steps:

(1) The mixture of glycol and the carbon nanofiller (in appropriate proportion) was passed through a calender to form a homogenous suspension (without agglomerates);

(2) The mixture of glycol and the carbon nanofiller was added to amorphous silica and mixed in appropriate proportion (know-how protected).

Table 1. The properties of carbon nanofillers.

Fillers	Density (g/cm^3)	Carbon Content (%)	Parameters (nm)	Specific Surface Area (m^2/g)
MWCNT	2.08	>90	9.5 (diameter) 1500 (length of a single tube)	300
CB	2.05	>95	11 (particle size)	350
GO	2.03	40	0.9 (distance between surfaces)	211
rGO	2.37	75	0.37 (distance between surfaces)	266

The shape of the carbon nanofillers was characterized using an HITACHI S3500 scanning electron microscope (SEM) (Krefeld, Germany) in SE mode.

The rheological properties were examined using an ARES rheometer (TA Instruments) (New Castle, DE, USA), equipped with two parallel plates (φ 25 mm) with a 0.3 mm gap between them. The size of the gap was optimized for all the fluids, based on previous research. All the viscosity measurements were performed at room temperature.

The STFs with the best rheological properties (highest maximal viscosity) were chosen for the impact test to study the force absorbing efficiency. The synthesized STFs were used for impregnation of the three-dimensional (3D) fabric (polyester three-dimensional woven fabric M8180 from Baltex Ltd. Warsaw, Poland), which was sealed between thin silicon membranes (130 × 30 × 15 mm). The impact tests were carried out using a drop tower, by dropping an impactor, with energy of 5 J, onto the sample (the procedure based on British Standard BS 7971–4:2002). The dependence between the force and time was registered by a force sensor.

The structural stability of the STFs was studied using a Turbiscan Lab analyzer (Formulation, Toulouse, France) with the light $\lambda = 880$ nm (see [27]). Some of the STFs samples were tested in a sealed cylindrical glass vial to show the changes in the arrangement of particles over time. The measurements were performed for 370 days at room temperature. The light intensity profiles were achieved with a scan step of 40 µm, along with the entire height of the measuring cell. The Turbiscan Lab has two synchronous detectors for the analysis of liquids (Figure 2). These devices detect the intensity of transmitted and backscattered light over the entire sample. Only part of the light passes through the sample, which is registered by sensors located on the walls of the chamber. The measurement was carried out without any mechanical or external stress, and therefore the true aging of the product was monitored. The resulting profile was plotted as the backscattered intensity of the light passing through the fluids versus the height of the dispersion. All the measurements were taken at room temperature. The peaks from the bottom of the vial and from the top of the unfilled vial were removed from the graph.

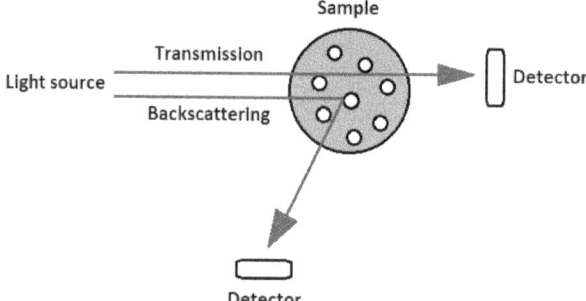

Figure 2. The principle of operation of the Turbiscan device.

3. Results and Discussion

3.1. Microstructure

The carbon additives for the STFs and amorphous silica were subjected to SEM observations using a HITACHI S3500 scanning electron microscope (SEM) in SE mode (Figures 3 and 4). It can be seen that all the particles used for testing have different shapes and various tendencies for agglomeration.

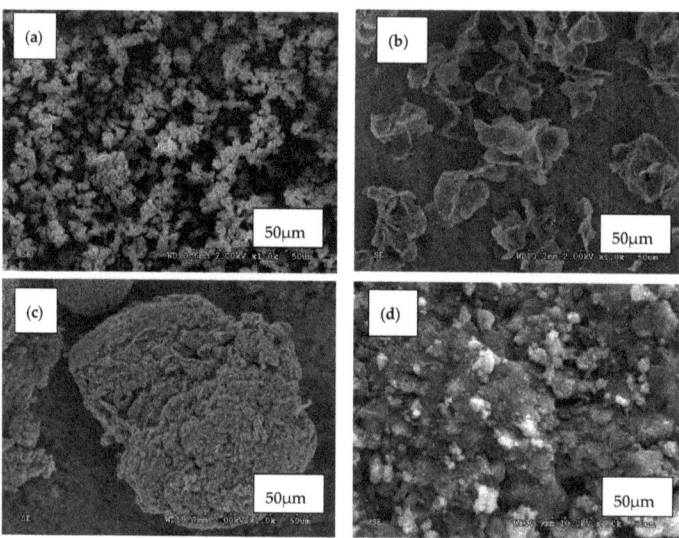

Figure 3. SEM images of (**a**) Graphene oxide GO; (**b**) Reduced graphene oxide (rGO); (**c**) Multi-walled carbon nanotubes (MWCNT); (**d**) Carbon black (CB) particles.

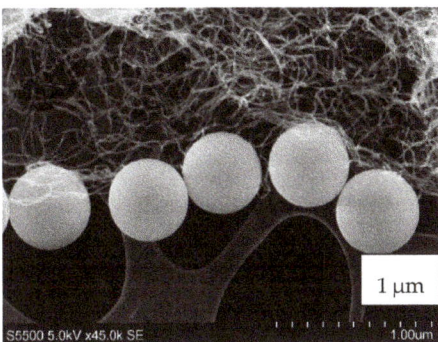

Figure 4. The shape and size of the MWCNT and spherical fumed silica KE-P50.

The particles of silica that were used for testing had a regular spherical shape with fairly uniform size distribution and low surface development (Figure 4). They had a slight tendency for agglomeration. The carbon fillers, in contrast, had a high tendency for agglomeration, forming bundles (especially MWCNT, Figure 3c). Graphene oxide and reduced graphene oxide exhibited flake shapes. Carbon black had an irregular shape also with the tendency for agglomeration.

In Figure 5, the distribution of silica and carbon nanofillers in the STFs are shown. The fluids with the highest carbon fillers content, 0.25 vol.%, were degassed (approx. 0.08 MPa), at 100 °C to determine the distribution of the solid components.

Mixing of fumed silica with carbon nanofillers in a carrier liquid leads to breaking the agglomerates of the fillers and as a result they are more evenly distributed among the silica particles [23,24,28]. This effect strongly depends on the filler used. The GO and rGO behave similarly, i.e., they are distributed between the silica particles (Figure 5a,b). Numerous agglomerates are also visible.

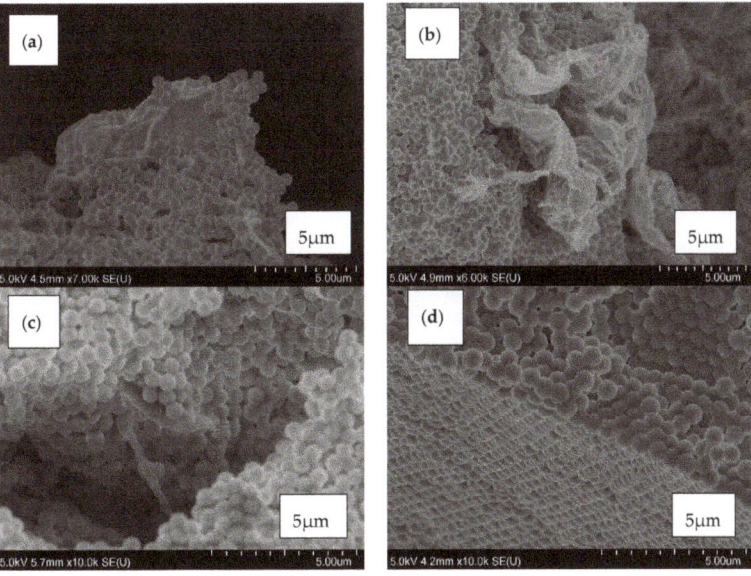

Figure 5. The arrangement of solid components in the shear thickening fluid (STF) with 55 vol.% of silica and 0.25 vol.% of (**a**) GO; (**b**) rGO; (**c**) MWCNT; (**d**) CB particles.

In Figure 5c, the distribution of silica and MWCNT in the STF is shown. The single tubes can be seen. Mixing of fumed silica with MWCNT leads to breaking of the agglomerates of the nanotubes and their even dispersion in the voids between the silica particles. The agglomerates of CB were destroyed (Figure 5d) and they are not visible in the image. The crushed CB particles could be placed between the silica spheres.

3.2. Rheological Properties

In Figure 6, the rheological properties of composite STFs with various additives are presented. When a drastic increase of the viscosity begins, the viscosity and the critical shear rate strongly depend on the type and content of the carbon nanofiller. Increasing the carbon fillers content moves the critical shear rate to lower values. It means that the shear thickening effect occurs more easily. The values of the zero shear viscosity (z-s-v), obtained for the suspensions, except for the STF with 0.25 vol.% of rGO, are similar (not higher than 70 Pa·s). With an increase of the shear rate, an abrupt increase of the viscosity is observed. The greatest effect of the dopant on the maximal viscosity is observed for the MWCNT. The highest values of the viscosity were obtained for 0.15 vol.%, 10,994 Pa·s, and for 0.25 vol.%, 12,213 Pa·s, with a critical shear rate below 2 s^{-1} for both fluids. The latter value is over five times higher than the result for the MWCNT-free fluid.

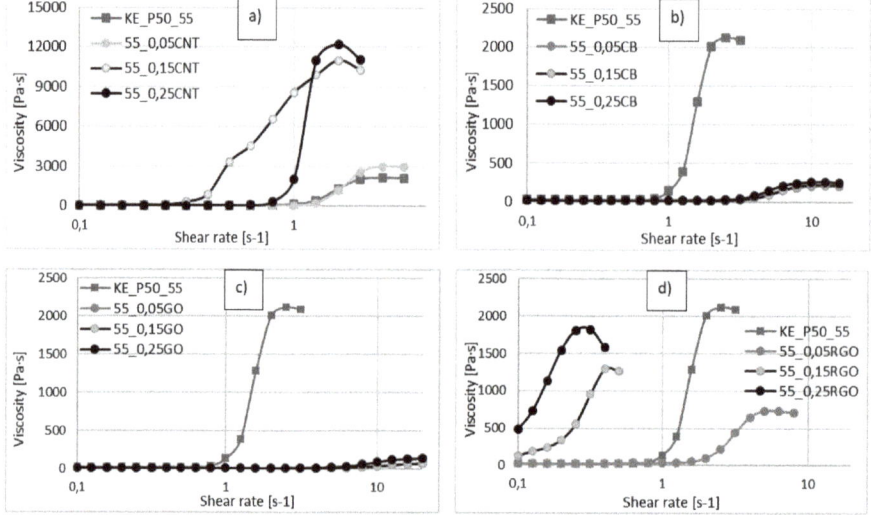

Figure 6. Viscosity versus shear rate for different volume fractions of carbon fillers. (**a**) CNT; (**b**) CB; (**c**) GO; (**d**) rGO), as compared with STF having 55 vol.% of KE-P50 without additives.

The lowest values of maximal viscosity were obtained for the samples with GO and CB (not higher than maximal viscosity for the pure STF, equal to 2127 Pa·s (Figure 6b,c). The viscosity values are low and the changes of the viscosity occur at higher shear rates, which can be explained by the fact that these additives react as a lubricant for the movement of the silica particles. The GO and CB cause the silica particles to move freely without being blocked. When the shear rate increases, the particles of CB and GO slightly retard the movement of silica particles, leading to a small thickening effect. However, this effect is not as strong as in the case of pure STF. Greater values were obtained for rGO (Figure 6d), however, the highest value of the maximal viscosity is still lower than the result achieved for the pure STF.

The rheological properties of the STFs depend on the ability of the silica particles to move against each other. This ability depends on the volume fraction of the solid components and their mechanical properties. The effect of a minor addition of carbon nanostructures is not fully clear. However, from the results presented, we can conclude that the addition of the carbon nanofillers increases the initial packaging of the solid components, and thus the (z-s-v) viscosity (GO, rGO). In the course of calendering and mixing, the structure of the carbon nanofillers becomes partially destroyed and located on the surface of silica particles and voids between them. Depending on its form, the carbon layer can act as either a lubricant or a friction increasing factor that retards the movement of the silica spheres. The latter behavior was observed for MWCNT. Apparently the MWCNT, due to their characteristic nanostructure and high mechanical properties, promote blocking of the silica particles movement. We assume that mixing the amorphous silica with MWCNT leads to breaking the agglomerates of the nanotubes. In the consequence of this process, the individual MWCNT can be dispersed in the voids between the silica particles, leading to increased friction between the particles and improved blocking of their movement.

3.3. Impact Absorption Ability

The STFs with 55 vol.% of amorphous silica and with the addition of carbon nanotubes were used for carrying out an absorption ability test. During the tests, none of the samples were destroyed.

Figure 7 presents the values of the absorbed force for STFs sealed in silicone form after one, two, and three strikes at the same point with the addition of MWCNT. One can see that a higher addition of MWCNT provides an increase of the absorbed force. The samples containing 0.25 vol% MWCNT show the highest absorbing properties (78% for the first strike). The lowest value of absorbed force, for the first strike, was obtained for STF without nanofillers (approximately 66%).

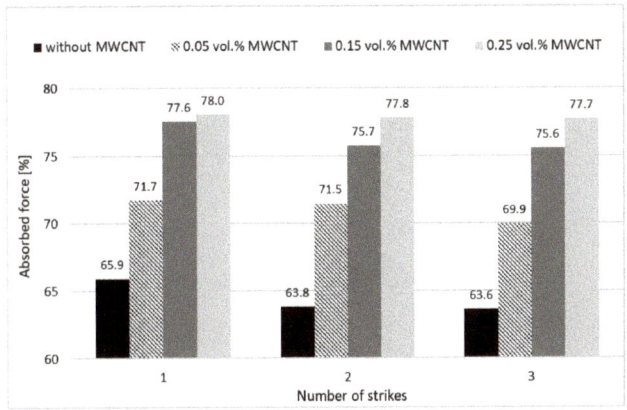

Figure 7. The percentage of absorbed force for STFs sealed in silicone forms after one, two, and three strikes at the same point, with various additions of MWCNT.

The mean value of the stability of the absorbed force after three strikes is satisfactory for all the samples. The changes after all the strikes (comparison between first and third strikes) are not greater than 3.5% for pure STF and approximately 0.5% for the sample with 0.25 vol.% of MWCNT. It means that the composite STFs provide high protection ability and also stability of the properties after three strikes at the same point. This fact is related to the specific mechanism of the protection, which is due to the reversible rearrangement of the solid particles within the STF structure.

3.4. Structural Stability

Usually, the published results for the stability studies of the STFs suspensions are based on optical observations of sedimentation in typical glass vials at room temperature [29]. However, such data are only comparative and do not allow for an accurate, quantitative comparison of the fluids. In the current study, the stability was measured using the Turbiscan LAB. The backscattered profiles were obtained for three fluids, i.e., one without carbon nanofillers and two with carbon additives. Because the values obtained for rGO, GO, and CB were similar, for the comparison, only the results for MWCNT and rGO are presented. The changes of the stability are presented in Figures 8–10. The spectra reflect the backscattered light flow in relation to the sample height (mm). The time of the measurement is represented by colors of the lines. Different colors of lines present destabilization of the structure of the STF. The blue color shows the beginning of the measurements and the red color shows the end of the measurement.

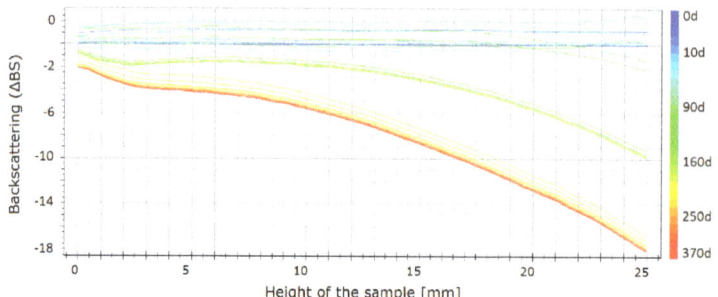

Figure 8. Backscattering for the STFs having 55 vol.% of KE-P50.

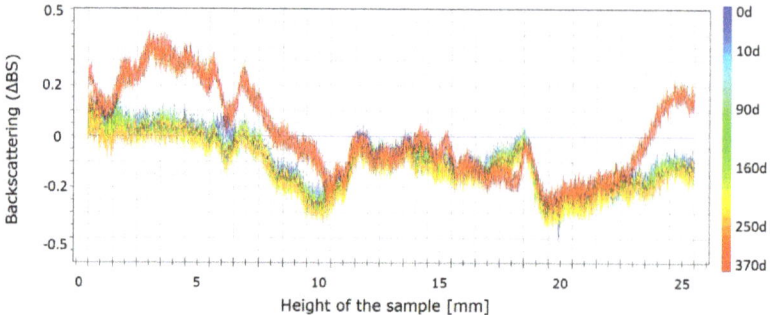

Figure 9. Backscattering for the STFs having 55 vol.% of KE-P50 and 0.25 vol.% of rGO within 370 days.

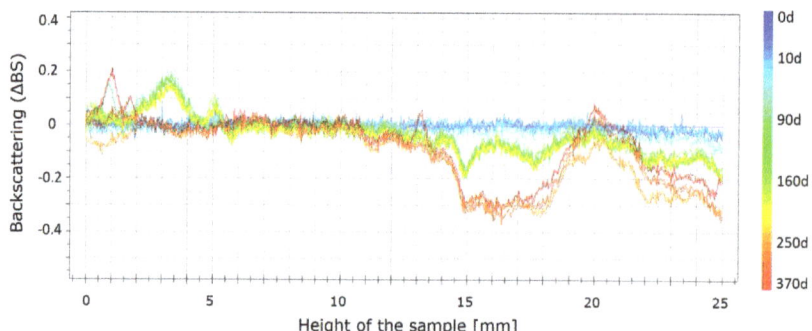

Figure 10. Backscattering for the STFs having 55 vol.% of KE-P50 and 0.25 vol.% of MWCNT within 370 days.

As shown in Figure 8, changes in the pure STF are visible. It should be mentioned that, after one year, a slight destabilization occurs through the entire vial. The mean values of backscattering are approximately 20, which means that the particles change their position slightly and as a result the properties of the STF change slightly. The values obtained for both samples with the carbon additives are shown in Figures 9 and 10. The absolute value of the change does not exceed one. It means that the changes are insignificant. For the sample with MWCNT, these changes are lower than for rGO. The changes are not higher than 0.4. It means that the addition of MWCNT provides the best structural stability of the fluids over time.

4. Conclusions

The composite STFs based on amorphous silica with polypropylene glycol and four types of carbon nanofillers were fabricated. Multi-walled carbon nanotubes, reduced graphene oxide, graphene oxide, and carbon black were applied as the carbon nanofillers. The microstructure, rheological properties and structural stability over time were studied and discussed. It was found that the MWCNT provide the most significant changes in the rheological properties of the STF, resulting in the highest maximal viscosity obtained in the rheological tests (almost five times greater than the result for the MWCNT-free fluid). We assume that, depending on the form of the carbon additive, it can act as either a lubricant or a friction increasing factor that retards the movement of the silica particles. Apparently the MWCNT, due to their characteristic nanostructure and high mechanical properties, increase the friction and promote blocking of the silica particles movement.

The carbon nanofillers lead to an increase of the structural stability of the STFs. The mean values of backscattering for pure STF are approximately 20 times higher than for STFs with carbon nanofillers, which means that STFs with carbon additives are more stable than the pure STFs.

The addition of the MWCNT also has a significant influence on the impact force absorbing efficiency. The silicone structure containing 55 and 0.25 vol.% of fumed silica and MWCNT, respectively, is able to absorb up to 78% of the impact force. Comparing with the structure with pure STF, this value is approximately 12% higher after the first strike.

The STFs containing MWCNT have great potential for application in smart protective structures, such as sport protective garments, liquid body armor, packaging systems and many other applications.

Author Contributions: Conceptualization, P.N.-D. and M.L.; methodology, software, analysis, P.N.-D., R.W., and M.P.; writing—original draft preparation, P.N.-D. and M.L.; writing—review and editing M.L. All authors have read and agreed to the published version of the manuscript.

Funding: The research was financed from the Subsidy Research Funds of the Warsaw University of Technology. The publication was financed from the "Excellence initiative – research university", program Open Science.

Conflicts of Interest: The authors declare no conflict of interest. The funders had no role in the design of the study; in the collection, analyses, or interpretation of data; in the writing of the manuscript, or in the decision to publish the results.

References

1. Young, D.F.; Munson, B.R.; Okiishi, T.H.; Huebsch, W.W. *A Brief Introduction to Fluid Mechanics*, 4th ed.; Wiley: Hoboken, NJ, USA, 2007.
2. Barnes, H.A. Shear-Thickening ("Dilatancy") in Suspensions of Nonaggregating Solid Particles Dispersed in Newtonian Liquids. *J. Rheol.* **1989**, *33*, 329–366. [CrossRef]
3. Popplewell, J.; Rosensweig, R.E. Magnetorheological fluid composites. *J. Phys. D Appl. Phys.* **1996**, *29*, 2297–2303. [CrossRef]
4. Pensini, E.; Rodriguez, B.; Marangoni, A.; Collier, C.; Elsayed, A.; Siwik, A. Shear Rheological Properties of Composite Fluids and Stability of Particle Suspensions: Potential Implications for Fracturing and Environmental Fluids. *Can. J. Chem. Eng.* **2019**, *97*, 2395–2407. [CrossRef]
5. Resiga, D.; Vékás, L. From high magnetization ferrofluids to nano-micro composite magnetorheological fluids: Properties and applications. *Rom. Rep. Phys.* **2018**, *70*, 501.
6. Chen, K.; Wang, Y.; Xuan, S.; Gong, X. A hybrid molecular dynamics study on the non-Newtonian rheological behaviors of shear thickening fluid. *J. Colloid Interface Sci.* **2017**, *497*, 378–384. [CrossRef]
7. Chhabra, R.; Richardson, J. *Non-Newtonian Flow in the Process Industries*; Butterworth-Heinemann: Oxford, UK, 1999.

8. Wagner, N.J.; Brady, J.F. Shear thickening in colloidal dispersions. *Phys. Today* **2009**, *62*, 27–32. [CrossRef]
9. Green, R.G.; Griskey, R.G. Rheological Behavior of Dilatant (Shear-Thickening) Fluids. Part II. Mechanism and Theory. *Trans. Soc. Rheol.* **1968**, *12*, 27–37. [CrossRef]
10. Das, S.; Riest, J.; Winkler, R.G.; Gompper, G.; Dhont, J.; Nägele, G. Clustering and dynamics of particles in dispersions with competing interactions: Theory and simulation. *Soft Matter* **2018**, *14*, 92–103. [CrossRef]
11. Majumdar, A.; Butola, B.S.; Srivastava, A. An analysis of deformation and energy absorption modes of shear thickening fluid treated Kevlar fabrics as soft body armour materials. *Mater. Des.* **2013**, *51*, 148–153. [CrossRef]
12. Lee, B.; Kim, I.-J.; Kim, C.-G. The Influence of the Particle Size of Silica on the Ballistic Performance of Fabrics Impregnated with Silica Colloidal Suspension. *J. Compos. Mater.* **2009**, *43*, 2679–2698. [CrossRef]
13. Wang, Y.; Chen, X.; Young, R.; Kinloch, I. A numerical and experimental analysis of the influence of crimp on ballistic impact response of woven fabrics. *Compos. Struct.* **2016**, *140*, 44–52. [CrossRef]
14. Lee, Y.S.; Wetzel, E.D.; Wagner, N.J. The ballistic impact characteristics of Kevlar® woven fabrics impregnated with a colloidal shear thickening fluid. *J. Mater. Sci.* **2003**, *38*, 2825–2833. [CrossRef]
15. Qin, J.; Zhang, G.; Zhou, L.; Li, J.; Shi, X. Dynamic/quasi-static stab-resistance and mechanical properties of soft body armour composites constructed from Kevlar fabrics and shear thickening fluids. *RSC Adv.* **2017**, *7*, 39803–39813. [CrossRef]
16. Tian, T.; Nakano, M.; Li, W. Applications of shear thickening fluids: A review. *Int. J. Hydromechatron.* **2018**, *1*, 238–257. [CrossRef]
17. Ge, J.; Tan, Z.; Li, W.; Zhang, H. The rheological properties of shear thickening fluid reinforced with SiC nanowires. *Results Phys.* **2017**, *7*, 3369–3372. [CrossRef]
18. Kuşhan, M.C.; Gürgen, S.; Ünalir, T.; Çevik, S. A novel approach for armor applications of shear thickening fluids in aviation and defense industry. *Int. Conf. Sci.* **2014**, *22*, 4.
19. Park, Y.; Kim, Y.; Baluch, A.H.; Kim, C.-G. Empirical study of the high velocity impact energy absorption characteristics of shear thickening fluid (STF). *Int. J. Impact Eng.* **2014**, *72*, 67–74. [CrossRef]
20. Zheng, S.; Xuan, S.; Jiang, W.; Gong, X.-L. Steady shear characteristic and behavior of magneto-thermo-elasticity of isotropic MR elastomers. *Smart Mater. Struct* **2015**, *24*, 85.
21. Wei, M.; Lv, Y.; Sun, L.; Sun, H. Rheological properties of multi-walled carbon nanotubes/silica shear thickening fluid suspensions. *Colloid Polym. Sci.* **2020**, *298*, 243–250. [CrossRef]
22. Kinloch, I.A.; Suhr, J.; Lou, J.; Young, R.J.; Ajayan, P.M. Composites with carbon nanotubes and graphene: An outlook. *Science* **2018**, *362*, 547–553. [CrossRef]
23. Hasanzadeh, M.; Mottaghitalab, V.; Babaei, H.; Rezaei, M. The influence of carbon nanotubes on quasi-static puncture resistance and yarn pull-out behavior of shear-thickening fluids (STFs) impregnated woven fabrics. *Compos. Part A Appl. Sci. Manuf.* **2016**, *88*, 263–271. [CrossRef]
24. Tan, Z.; Ma, H.; Zhou, H.; Han, X.; Cho, C. The influence of graphene on the dynamic mechanical behaviour of shear thickening fluids. *Adv. Powder Technol.* **2019**, *30*, 2416–2421. [CrossRef]
25. Głuszek, M.; Żurowski, R.A.; Kubiś, M.; Wisniewski, T.S.; Szafran, M. Shear thickening behavior and thermal properties of nanofluids with graphite fillers. *Mater. Res. Express* **2018**, *6*, 015701. [CrossRef]
26. Nakonieczna, P.; Wierzbicki, Ł.; Leonowicz, M.; Lisiecki, J.; Nowakowski, D. Composites with impact absorption ability based on shear thickening fluids and auxetic foams. *Compos. Theory Pract.* **2017**, *17*, 67–72.
27. Nakonieczna, P.; Wojnarowicz, J.; Wierzbicki, L.; Leonowicz, M. Rheological properties and stability of shear thickening fluids based on silica and polypropylene glycol. *Mater. Res. Express* **2019**, *6*, 115702. [CrossRef]
28. Neelanchali, A.; Chouhanand, H.; Bhatnagar, N. Synthesis of Shear Thickening Fluids for Liquid Armour Applications. *J. Manuf. Sci.* **2016**, *6*, 9–17.
29. Wierzbicki, Ł.; Danelska, A.; Olszewska, K.; Tryznowski, M.; Zielińska, D.; Kucińska, I.; Leonowicz, M. Shear thickening fluids based on nanosized silica suspensions for advanced body armour. *Compos. Theory Pract.* **2013**, *4*, 241–244.

© 2020 by the authors. Licensee MDPI, Basel, Switzerland. This article is an open access article distributed under the terms and conditions of the Creative Commons Attribution (CC BY) license (http://creativecommons.org/licenses/by/4.0/).

Article

Relationship between Viscosity, Microstructure and Electrical Conductivity in Copolyamide Hot Melt Adhesives Containing Carbon Nanotubes

Paulina Latko-Durałek [1,2,*], Rafał Kozera [1], Jan Macutkevič [3], Kamil Dydek [1] and Anna Boczkowska [1,2]

[1] Faculty of Materials Science and Engineering, Warsaw University of Technology, 02-507 Warsaw, Poland; rafal.kozera@pw.edu.pl (R.K.); kamil.dydek@pw.edu.pl (K.D.); anna.boczkowska@pw.edu.pl (A.B.)
[2] Technology Partners Foundation, 02-106 Warsaw, Poland
[3] Faculty of Physics, Vilnius University, 10222 Vilnius, Lithuania; jan.macutkevic@gmail.com
* Correspondence: paulina.latko@pw.edu.pl

Received: 10 September 2020; Accepted: 6 October 2020; Published: 9 October 2020

Abstract: The polymeric adhesive used for the bonding of thermoplastic and thermoset composites forms an insulating layer which causes a real problem for lightning strike protection. In order to make that interlayer electrically conductive, we studied a new group of electrically conductive adhesives based on hot melt copolyamides and multi-walled carbon nanotubes fabricated by the extrusion method. The purpose of this work was to test four types of hot melts to determine the effect of their viscosity on the dispersion of 7 wt % multi-walled carbon nanotubes and electrical conductivity. It was found that the dispersion of multi-walled carbon nanotubes, understood as the amount of the agglomerates in the copolyamide matrix, is not dependent on the level of the viscosity of the polymer. However, the electrical conductivity, analyzed by four-probe method and dielectric spectroscopy, increases when the number of carbon nanotube agglomerates decreases, with the highest value achieved being 0.67 S/m. The inclusion of 7 wt % multi-walled carbon nanotubes into each copolyamide improved their thermal stability and changed their melting points by only a few degrees. The addition of carbon nanotubes makes the adhesive's surface more hydrophilic or hydrophobic depending on the type of copolyamide used.

Keywords: carbon nanotubes; hot melt adhesives; electrical conductivity; viscosity; microstructure

1. Introduction

Adhesive bonding is frequently used to join and repair lightweight thermoplastic or thermosetting matrix composite parts in the automotive and aircraft industry since it eliminates rivets, thus lowering the stress concentration and total weight of the final parts [1–3]. Nevertheless, the separation of two bonded composites by an electrically insulating layer formed by the polymeric adhesive is a real problem in terms of the lightning strike protection of the aircraft or automotive composite structures. Therefore, the idea is to use electrically conductive adhesives (ECAs) which strongly bond composites together while at the same time providing an electrical interconnection between them [4]. These double functions of ECAs are also desired in the electronics industry (interconnection of chips on printed circuit boards) and in the photovoltaic industry (assembly of the aluminum back surface field or shingled solar cells) and make them a promising solution for the replacement the traditional Pb–Sn solder alloys, much heavier than ECAs [5,6].

ECAs consist of a polymer matrix and electrically conductive filler or nanofiller. Depending on the type of filler and its concentration, ECAs can be divided into isotropic, anisotropic and non-conductive according to the percolation theory. Isotropic ECAs have a high content of conductive filler that exceeds the percolation threshold and they are able to conduct current in all directions (x, y, z), unlike anisotropic and non-conductive ECAs which are conductive in only one direction [7]. Silver in the form of flakes [8], powder [9], nanowires [10] or dendrites [11] is the main type of electrically conductive filler which for a long time has been used in ECAs due to its excellent electrical conductivity and thermal stability. However, the minimum content of silver necessary to achieve a sufficient level of electrical conductivity at which a decrease in polymer resistivity is observable, varies from 25 wt % up to even 80 wt % [12]. A higher amount of filler leads to a significant weight increase, increases the price of the adhesive and lowers the mechanical properties [13]. Hence, the research focused on decreasing the silver percentage by partially replacing it with alternative electrically conductive nanofillers or using these nanofillers alone. Studies frequently focus on such promising materials as carbon-based nanofillers, e.g., graphene, single-walled carbon nanotubes (SWCNTs), multi-walled carbon nanotubes (MWCNTs), reduced graphene oxide or carbon nanofibers, known to be highly conductive and lightweight. The amount of these nanofillers required for electrically conductive network formation, identified by percolation threshold, is much lower in comparison to the metallic particles, usually between 1 and 3 wt % [14]. Compared to other types of carbon nanofillers, MWCNTs are characterized by excellent electrical properties ($8 \times 10^{-6} \div 20 \times 10^{-6}$ Ωm), thermal conductivity (λ = 6600 W/mK at 100 K), a high Young's modulus (1.7–2.4 TPa) and tensile strength (100 GPa) as well as low cost (100 EUR/1 kg) [15,16].

The polymers commonly used as the matrix for ECAs are those based on thermosets such as acrylic, epoxy, urethane, cyanoacrylate or silicone available as one- or two-component liquid systems. Epoxy resin, the most popular matrix for the adhesive, has been doped with various types of carbon nanofillers using the three-roll mill technique by Lopes et al. [6]. The greatest decrease in epoxy resistivity—achieved for SWCNTs, MWCNTs and exfoliated graphite—was higher than that of graphite, carbon fiber and nanofibers. In other work, epoxy resin was mixed with MWCNTs and the highest electrical conductivity was 10^0 S/m, much lower than that obtained at 80 wt % silver [17]. Enhancing the electrical conductivity of the ECAs containing carbon nanofillers is realized by increasing their content up to even 50 wt % reported for reduced graphene oxide (r-GO) in epoxy resin applying ultrasonic technique [18]. The highest achieved volume electrical conductivity was 3.4×10^{-8} S/m, or two orders of magnitude lower than that determined for the composites containing only 2 wt % of graphene [19]. It is associated with the strong effect of the type of carbon nanofiller on the electrical conductivity, mainly its aspect ratio, purity, surface area and the presence of functional groups described previously for many thermosets and thermoplastic polymers [17,20]. Furthermore, it was shown that the viscosity, crystallinity content, polarity of the polymer as well as the mixing technique and applied conditions affect the dispersion and distribution of the nanofillers in the polymer matrix. Carbon nanofillers are synthesized in the form of strongly connected agglomerates which must be destroyed during processing to a homogenous dispersed state. The remaining agglomerates, and the orientation and alignment of the dispersed nanofiller are responsible for the level of electrical conductivity achieved in the polymer composites [21,22].

In order to increase the electrical conductivity, carbon nanofillers can be modified by metallization with silver, copper, or nickel particles by chemical reactions. Acrylic-based adhesives containing 2 wt % MWCNT metallized with silver resulted in a volume electrical conductivity of about 2.8×10^{-6} S/m [23], an insufficient value for application as ECAs. Therefore, considerable effort is being made put to form hybrid ECAs by mixing micro and nanofillers together due to their confirmed synergic effect caused by changing the contact resistance inside the electrical network [24]. Marcq and co-workers [25] found that epoxy adhesives doped with silver and SWCNTs, DWCNTs (double-walled carbon nanotubes) and MWCNTs do indeed possess higher electrical conductivity than adhesives containing only 25 wt % silver flakes. Similar improvement of the electrical conductivity was described for epoxy adhesives mixed with micron silver flakes, nano silver spheres and treated CNTs in comparison to epoxy resin containing only silver flakes [26]. A synergic effect occurring between fillers with different morphologies was also described for acrylate resin mixed with silver-plated graphene, leading to the greatest electrical conductivity of 4.8×10^2 S/m at 40 wt % content of the hybrid filler [27].

The fabrication of ECAs from thermosets, especially with the hybrid nanofillers, involves many steps, chemical reactions, hazardous solvents and their curing process requiring the addition of catalysts or initiators, UV light or heat which make the process complicated and lengthy [28]. Besides, the incorporation of carbon nanofillers is limited to 3 wt % by an extremely high increase in the viscosity which makes the manufacturing process difficult and affects the quality of the composites [29]. Therefore, it seems much easier to use thermoplastic non-reactive hot-melt adhesives (HMAs), solvent-free and environmentally friendly polymeric materials available in a solid state in the form of pellets, ropes, sticks or blocks. They can be applied to the surface by using a hot gun or in the form of thin films or fabrics. When HMAs are melted at a temperature above their melting point (usually in the range of 150–200 °C) they become a molten liquid which solidifies at room temperature. Because HMAs thicken rapidly, they are frequently used in those processes where manufacturing time is important and solvents are not desired. Moreover, they offer a high mechanical strength to the substrate without the need for special surface preparation. In addition, commercial HMAs offer a high range of properties together with fairly low prices, safe storage and operation. In comparison to the liquid adhesives like epoxy or urethanes which become solidified via curing or crosslinking reaction, HMAs do not form three-dimensional structures when they solidify and can thus be termed as non-structural adhesives. Nowadays, they are used in the packaging industry (bottle labeling, tapes, carton sealing), the automotive industry to bond plastic parts (rear tail gate, bumper, aesthetic skin), wood-based materials (parquet floor), book binding, temporary attachment (coupons, instructions), shoe manufacturing (insole bonding, tongue fixing), textile bonding (diapers, napkins) or as wearable electronic devices and lightweight constructions [2,30,31].

HMAs are complex in composition, therefore they cannot be classified as typical thermoplastic polymer but rather as blends. In our previous work, we attempted to identify all the components of the copolyamide-based HMAs using a nuclear magnetic resonance but we could ascertain their structure only partially [32]. The main component in HMAs is the thermoplastic polymer which solidifies upon cooling and provides the mechanical strength of the bond. Homopolymers or copolymers such as polyolefins, ethylene-vinyl acetate copolymers (EVA), polyurethanes, polyamides, copolyamides, styrene block copolymers (SIS, SEBS) or polyesters are mainly used in HMAs. Low molecular weight resin or tackifiers control the viscosity of the system, adjust its glass transition temperature and provide the adhesive properties. Waxes can decrease the viscosity and enhance the crystallization rate resulting in higher setting speed. Finally, there are also several other specific fillers and modifiers that can be used for the improvement of thermal stability and shelf life [31,33,34]. The doping of thermoplastic HMAs with electrically conductive fillers or nanofillers for their application as ECAs is less popular than for thermosets. Polyurethane-based HMAs containing MWCNTs or graphene were extensively studied by Santamaria and co-workers [35–39]. Composites were prepared by mixing HMA powder with up to 6 wt % of MWCNTs or graphene using a small scale twin screw extruder. They reported that the addition of MWCNTs or graphene did not significantly change properties such as the melting point, viscosity, crystallization or tackiness which could limit their use as ECAs. The analyzed electrical conductivity of the adhesives after cooling reached the value of 6×10^{-2} S/m which is higher than that described in the literature for ECAs based on thermosets. Similar electrical properties (10^{-2} S/m) were found for polyolefin HMAs mixed with 5 wt % MWCNT using the same direct melt–mixing approach [40]. Authors found that the integration of MWCNTs in polyolefin HMAs resulted in the improved adhesion of the bonded joints, however, the significant increase in melt viscosity made it impossible to apply adhesives containing more than 3 wt % MWCNT. Cecen et al. examined silver-coated wollastonite fibers as a conductive filler for the EVA copolymer [41]. The percolation threshold was found at 8 vol% of the filler, much higher than for the carbon nanofiller, and at 29 vol% the electrical conductivity reached a value of 1.8×10^5 S/m. Unfortunately, such high content of the conductive filler significantly decreased the adhesive properties in comparison to pure EVA. In the other works, HMA based on EVA was mixed with polypyrrole as a conductive filler which at 30 vol% resulted in the electrical conductivity of app. 1×10^2 S/m and a 15–20% of improvement of the adhesive properties [42].

The aim of this study was to characterize the new type of the ECAs based on copolyamide HMAs and MWCNTs. Since HMAs are complex in their structure, the idea was to test four types of these copolyamides to analyze the effect of their melt viscosity on the dispersion and distribution of MWCNTs. Previously, we analyzed the percolation threshold in two types of copolyamide HMAs and found it to be below 3 wt % [32]. Therefore, in that work, copolyamides were doped with 7 wt % of MWCNTs using a half-industrial extruder machine allowing to obtain the percolated network. The examination of their electrical, thermal and adhesive properties allowed for finding the relationship between the viscosity of pure copolyamides, the state of MWCNT dispersion obtained and the properties of the final HMAs containing electrically conductive nanofiller.

2. Materials and Methods

For this study, 4 types of thermoplastic copolyamides (coPAs) belonging to the group of HMAs were provided by EMS Griltech from Switzerland. According to the producer, they consisted of randomly arranged segments of PA6 and PA66 and differed in their properties as shown in Table 1. The conductive nanofiller used was MWCNTs with the trade name NC7000 from Nanocyl, Sambreville, Belgium) synthesized by catalytic carbon vapor deposition process. The average diameter of a MWCNT is 9.5 nm, length 1.5 µm and with purity >95%. All coPAs were mixed with 7 wt % of MWCNT using a half industrial line using a twin co-rotating screw industrial extruder by Nanocyl under the same processing conditions: an extrusion temperature of 200 °C and a rotational speed of 200 rpm. The neat coPAs and their masterbatches were dried in a vacuum oven at 80 °C for 12 h before further processing.

Table 1. Properties of thermoplastic copolyamides (coPAs) used in the study.

Designation	Trade Name	Melt Viscosity 160 °C/2.16 kg (Pa·s)	Melt Volume Rate 160 °C/2.16 kg	Melting Point (°C)
coPA1	Griltex® 1330	1200	9	125–135
coPA2	Griltex® 2A	600	18	120–130
coPA3	Griltex® 1858	350	30	110–120
coPA4	Griltex® 1566	150	70	115–125

Rheology measurement was performed on an ARES rheometer (Rheometric Scientific Inc., TA Instruments, New Castle, DE, USA) using a parallel plate geometry. The samples with a diameter of 1.5 mm and a thickness of 2 mm were prepared using a HAAKE™ Mini Jet Pro Piston Injection Molding System (ThermoScientific, Karlsruhe, Germany). Firstly, the linear elastic range was determined by conducting the amplitude sweep test of the materials. From the obtained graph, the amplitude strain was chosen as the highest value just before the moduli decreasing. Afterwards, the stress-controlled dynamic oscillatory test of neat coPAs and their masterbatches with MWCNTs was performed at 180 °C with a frequency sweep in the range of 0.1–100 Hz.

The macrodispersion of MWCNTs in the masterbatches was analyzed using a light transition microscope (Biolar-PL, Polskie Zakłady Optyczne, Warsaw, Poland). Samples for the test in the form of slides with a thickness of 2–3 µm were cut directly from the masterbatch pellets using an ultramicrotome (EM UC6, Leica, Vienna, Austria). The macrodispersion of MWCNTs was analyzed from several micrographs with image software (ImageJ version 1.52a) by the exclusion of those agglomerates with a diameter lower than 1 µm. Area ratio (A_A) understood as the percentage of MWCNT agglomerates was calculated by dividing the area of all agglomerates by the total area.

Characterization of MWCNT dispersion and arrangement in nanometer scale was analyzed using a transmission scanning electron microscope (HR STEM S5500, Hitachi, Krefeld, Germany) at the voltage of 30 kV. For this, thin slides of 80–90 nm were cut directly from the masterbatch pellets using an ultramicrotome (EM UC6, Leica, Vienna, Austria). The cutting process was carried out with diamond knives that are suitable for trimming and sectioning, at a temperature of 100 °C and with a cutting speed of 1 mm/s.

Broadband dielectric spectroscopy of the studied materials was performed at room temperature using a LCR HP4284A meter (Keysight Technologies, Santa Rosa, CA, USA). The equivalent electrical circuit was selected as the capacitance and the tangent of losses connected in parallel. From these quantities, according to the planar capacitor formula, the complex dielectric permittivity ε^* of the studied materials was calculated. The relation $\sigma = 2\pi\nu\varepsilon_0\varepsilon^*$ was used for the determination the electrical conductivity σ, where $\omega = 2\pi\nu$ and ν is the frequency of electromagnetic waves. To minimize the contact resistance silver paste was applied and the amplitude of the electric field was chosen as 1 V. Measurements under such conditions make it possible to reach the best signal to noise ratio in comparison with measurements at lower voltages and to avoid the nonlinear effects observed at higher voltages.

The volume DC electrical conductivity of the masterbatches containing MWCNTs was carried using the Keithley 6221/2182A measuring set (Cleveland, OH, USA), equipped with copper electrodes. Samples for the test were produced by thermo-pressing the dried pellets into rectangular shapes with the dimensions of 70 mm × 10 mm and thickness of around 1.5 mm. For each masterbatch, 5 samples were tested within the current range from 1 to 200 nA. To compensate for thermoelectric effects, measurements were made in the delta mode, using the four-point method.

Thermal stability of the materials expressed by the degradation temperature occurs at 2% ($T_{2\%}$) and 5% ($T_{5\%}$) weight loss, and by maximum peak (T_d) was designated by thermogravimetric analysis (TGA) using a TGA Q500 (TA Instruments, New Castle, DE, USA). For that 10 ± 0.2 mg samples were placed in an aluminum crucible and heated from 0 °C to 1000 °C with a heating rate of 10 °C/min. The atmosphere was nitrogen with a flow rate from 10 to 90 mL/min.

Thermal properties of all materials were examined using a Q1000 Differential Scanning Calorimeter (TA Instruments, New Castle, DE, USA) by placing the samples with a weight of 8.5 ± 0.2 mg in an aluminum hermetic pan under a nitrogen atmosphere. The applied program was heat–cool–heat from −60 °C to 250 °C with a scan rate 10 °C/min and the obtained curves were analyzed using TA Universal Analysis 2000 software version 4.5A. The glass transition temperature (T_g), melting point (T_m) and enthalpy of melting (ΔH_m) were determined from the heating curves, while the crystallization temperature (T_c) and the enthalpy of crystallization (ΔH_c) were found in the cooling curve. Due to the lack of data on the enthalpy of melting of 100% crystalline coPAs, the crystallinity content was not calculated.

Wettability properties (static contact angle (SCA), surface energy (SE)) were measured by a contact angle measurement system (Data Physics GmbH OCA 15, Filderstadt, Germany). All angles of each sample were measured at least three times across the sample surface using the sessile drop method, by dispensing 3 µL (SCA), of deionized water on the sample's surface. The round samples for the wettability test were made by a HAAKE™ Mini Jet Pro Piston Injection Molding System (ThermoScientific, Karlsruhe, Germany) with a diameter of 1.5 mm and thickness of 2 mm, the same as for the rheological test. For both pure coPAs and their masterbatches, the injection parameters were as follows: barrel temperature—220 °C, mold temperature—40 °C, injection pressure—600 bar, injection time—8 s, post pressure—500 bar and post-pressure time—8 s.

3. Results and Discussion

HMAs are characterized by their application temperature which for the selected coPAs is between 160 °C and 220 °C, therefore the rheological test was performed at 180 °C. Unfilled coPAs differ in viscosity which is the lowest for coPA4, followed by coPA3, coPA2 and the highest for coPA1, as presented in Figure 1a. All coPAs behave as Newtonian liquids since the frequency applied has no effect on the complex viscosity [43]. Unlike coPAs, masterbatches which exhibit a strong shear thinning behavior causing a decrease in their viscosity have been described for many thermoplastic polymers filled with CNTs [29]. In comparison to neat coPAs, the viscosity increases by about 4–5 orders of magnitude for the low viscosity coPA3 and coPA4 and about three orders of magnitude for the more viscous coPA1 and coPA2. The analysis of the storage (G′) and loss (G″) modulus provides information about the elastic and viscous properties of polymers, respectively [44]. Figure 1b,c show the sharp growth of G′ and G″ as the frequency increases. Because G″ is higher than G′, coPAs behave more like viscous liquid than elastic. The character of both modulus curves changes in the presence of 7 wt % MWCNT in an almost linear manner across the whole frequency range. For the studied coPA masterbatches the effect of MWCNT addition is visible through an increase of about 5–6 and 3–4 orders of magnitude for G′ and G″, respectively. The jump in G′ and G″ is caused by the interaction between the polymer macromolecules and CNTs and disturbing the macromolecule chains' movement. Since the MWCNTs used do not contain any functional groups which promote the formation of covalent bonds with the polymer chains, MWCNTs are connected with coPA chains by van der Waals forces [45]. However, this scenario is favored in low viscosity polymers because MWCNTs can enter easily between the macromolecule chains. The opposite is true in more viscous coPAs, where MWCNT dispersal is hampered due to the high degree of chain entanglement. Therefore, more hydrogen bonds are formed between the polymer chains themselves than between the polymer and MWCNTs [46]. Hence, the differences in the MWCNT dispersion should be expected. Moreover, it seems that the highest compatibility expressed by the value of G′ and G″ occurs for the coPA3 matrix. Rheological analysis confirms that these types of HMAs are a good polymer matrix for MWCNTs because there is a clear change in the viscoelastic properties of coPAs, and at high concentration such as 7 wt %, the nanofiller forms a percolated structure. The obtained values of both moduli (~10^5 Pa) are the same as those determined for HMAs based on polyurethane [39].

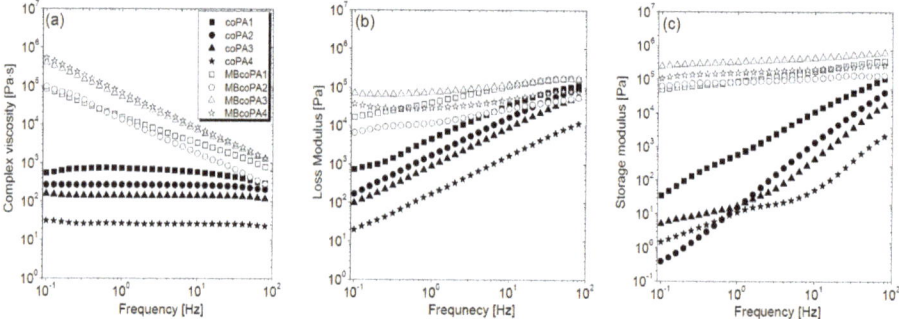

Figure 1. (a) Complex viscosity; (b) loss modulus G″ and (c) storage modulus G′ in the frequency dependence for the neat coPAs and their masterbatches containing 7 wt % multi-walled carbon nanotube (MWCNT).

In order to fully realize the potential of CNTs, the key step is to disperse the nanofiller uniformly in the polymer matrix. One of the methods of achieving this is twin screw extrusion where due to high shear force the pristine MWCNT agglomerates are effectively broken [47]. On the one hand, the less viscous polymer allows for a better CNT dispersion due to the easy infiltration of the nanofillers, whereas when the polymer is more viscous, a higher shear force occurs during extrusion, leading to better CNT dispersion [48]. To analyze the effect of the coPA melt viscosity on the MWCNT dispersion, four types of coPAs were mixed with 7 wt % MWCNT under the same extrusion conditions. Table 2 contains the micrographs of the masterbatches where black dots signify MWCNT agglomerates quantitatively expressed by their area ratio A_A (last column). It is seen that there is no linear dependence between the viscosity of coPAs and the number of MWCNT agglomerates. This is because the fewest agglomerates were found in coPA3, which has a medium viscosity. Indeed, for coPA1, which is the most viscous, the percentage ratio of agglomerates is lower than for coPA2 and coPA3, characterized by lower viscosity. This is consistent with the theory about the positive effect of the high shear force on the CNT agglomerate breakage, also confirmed by the lowest diameters (<40 μm) of MWCNT agglomerates in coPA1 (see column 3, Table 2). Conversely, in coPA3, where A_A is the lowest, agglomerates have higher diameters because the shear force during extrusion is lower than in the case of coPA1. Despite the significant difference in the melt viscosity, masterbatches of coPA2 and coPA4 have similar A_A and agglomerate diameters. The analysis of the nanofiller macrodispersion for coPA masterbatches showed that they are uniformly distributed in the polymer matrix. The homogenous distribution of MWCNTs in each coPA was also confirmed by the images given by a high-resolution microscope (Figure 2). From the presented micrographs it is shown that MWCNTs were not arranged or oriented in any specific direction.

Table 2. The comparison of MWCNTs' macrodispersion in the used coPAs expressed by agglomerates diameters and the area ratio. MV = melt viscosity in (Pa·s); s = standard deviation.

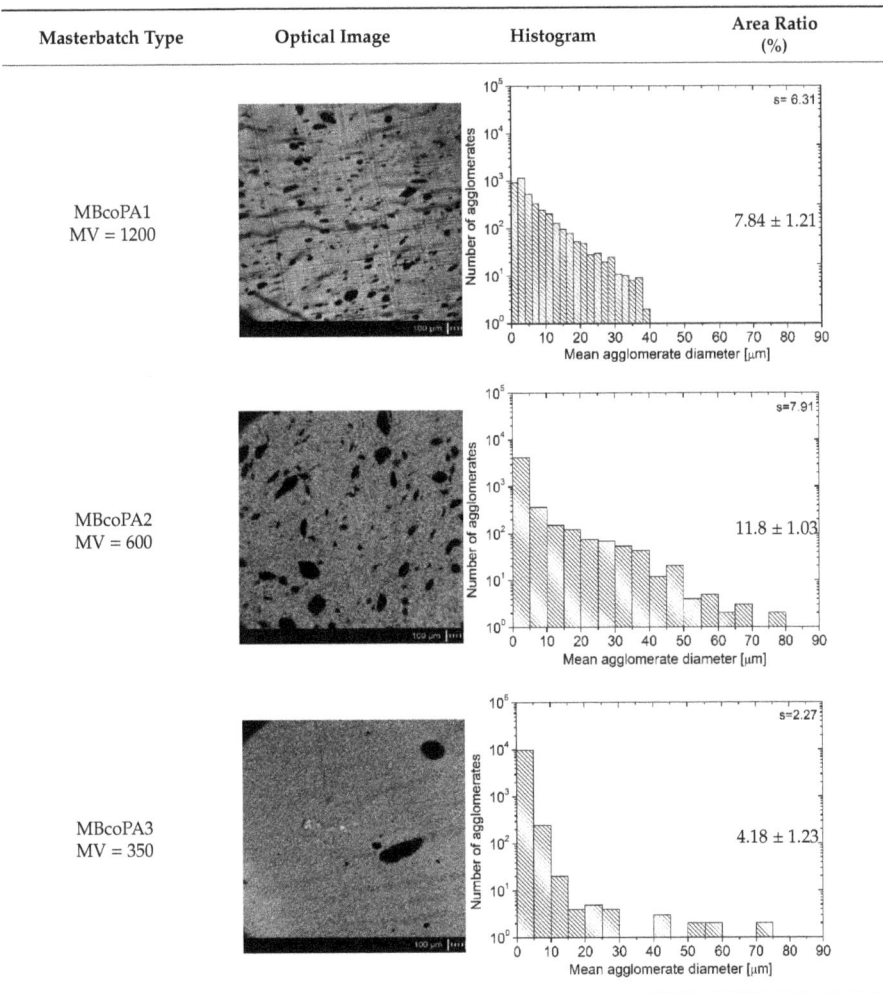

Table 2. *Cont.*

Masterbatch Type	Optical Image	Histogram	Area Ratio (%)
MBcoPA4 MV = 150			11.6 ± 1.14

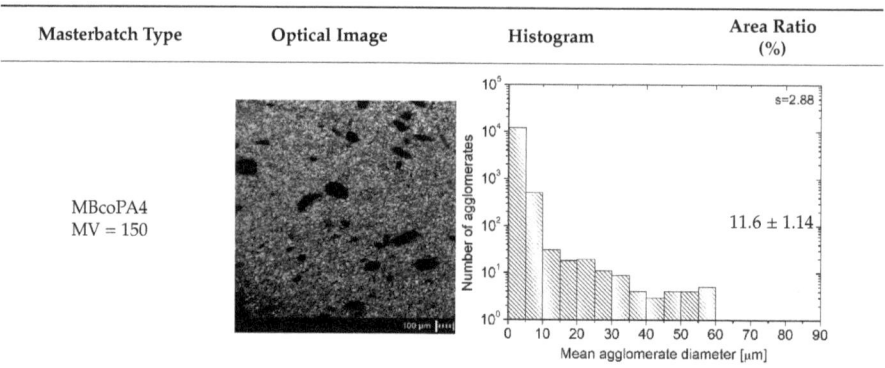

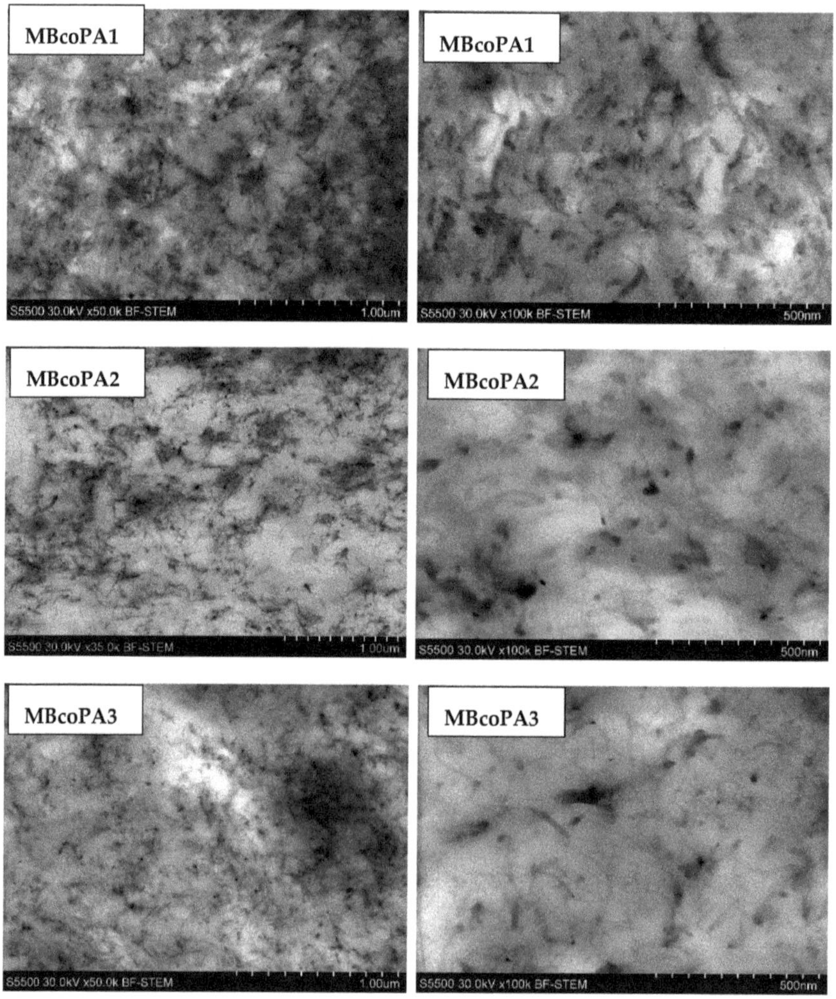

Figure 2. *Cont.*

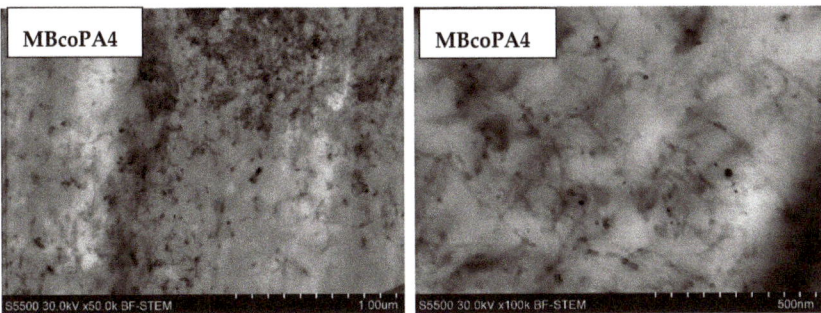

Figure 2. HR-STEM images of the microstructure of the copolyamide masterbatches containing 7 wt % MWCNT.

The electrical conductivity of the coPA HMAs is the main property for their final application as ECAs. It is known that only a uniform dispersion and distribution of nanofillers such as MWCNTs results in high electrical properties. Figure 3 presents the average values of the volume electrical conductivity for four types of masterbatches, with the highest recorded for MBcoPA4, followed by MBcoPa1, MBcoPA4 and the lowest for MBcoPA2. These results were correlated with the calculated A_A and it is clearly shown that electrical conductivity increases when A_A decreases. This means that MBcoPA3 possessed the highest electrical conductivity equal to 0.67 S/m because the percentage ratio of MWCNT agglomerates was the smallest (4.18%). The values of electrical volume conductivity achieved for MBcoPA3 are higher than those reported for the other types of ECAs containing carbon nanofillers but lower than those for ECAs with silver flakes, as listed in Table 3. Despite this, the weight of the carbon nanofillers used was less than that of the metal fillers, which is highly desired in ECAs. Assuming the price of coPA to be 10 EUR/1 kg, the price of MWCNTs to be 100 EUR/1 kg and the price of the silver flakes, 1000 EUR/1 kg, the cost of the filler in coPA adhesive will be 43% for MWCNTs and almost 99% for silver.

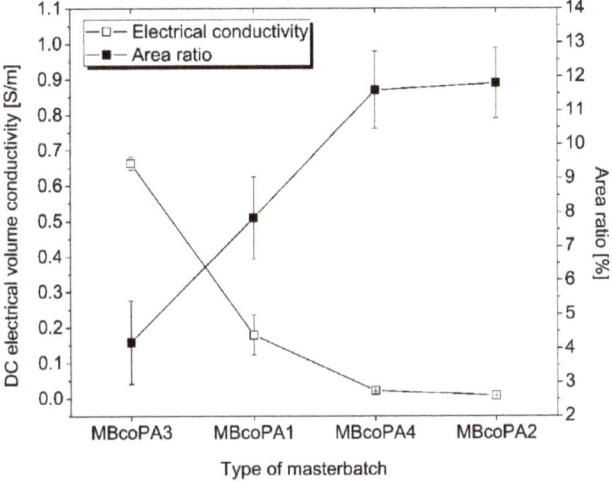

Figure 3. Dependence between the volume electrical conductivity and A_A for the studied masterbatches.

MBcoPA1 and MBcoPA3 were about one order of magnitude higher than for polyolefin HMAs containing 5 wt % of MWCNTs [49] and polyurethane ECAs mixed with 6 wt % of graphene [39]. In Table 3, there is a comparison of the electrical conductivity of the different ECAs.

Table 3. Electrical conductivity in different types of electrically conductive adhesives (ECAs).

Adhesive Matrix	Filler Type	Filler Content (wt %)	Electrical Conductivity (S/m)	Ref.
epoxy	silver flakes	70	10^2	[11]
epoxy	reduced graphene oxide	50	10^{-8}	[18]
epoxy	MWCNT	12	10^{-1}	[17]
ethylene-vinyl acetate	graphite nanoplatelets	30	10^{-5}	[50]
polyurethane HMA	graphene	6	10^{-2}	[39]
polyolefin HMA	MWCNT	5	10^{-2}	[40]
coPA3 HMA	MWCNT	7	0.67	this work

The frequency dependence of AC electrical conductivity and dielectric permittivity for pure coPAs and their masterbatches containing 7 wt % MWCNT is shown in Figure 4a,b. As expected, pure coPAs have rather low (around 10^{-8} S/m) electrical conductivity at low frequency and even 10^{-5} S/m at high frequency. Moreover, no frequency independent (DC) conductivity is observed in the conductivity spectra of pure coPAs, therefore the AC electrical conductivity is related to some dielectric relaxation rather than the electrical transport [50]. The dielectric permittivity (ε') is around 10 at the whole frequency range and it is much higher than that reported for the typical homopolymers like PA6 and PA66 [51]. Such discrepancies as well as the slight differences in the dielectric properties presented for unfilled coPAs may be associated with the HMA complex formulation (resin, tackifier, wax, etc.) affecting their polarity. The incorporation of 7 wt % MWCNT into coPAs resulted in a significant increase in the electrical properties and dielectric permittivity observed for each masterbatch. The electrical conductivity rose by 3 to 6 orders of magnitude depending on the type of coPA, up to 10^{-2} S/m. At low frequencies, conductivity curves have a plateau that corresponds to the DC electrical conductivity of the electrically percolated network [14]. Determined at 129 Hz, electrical conductivity was the lowest for MBcoPA2 (8.02×10^{-6} S/m), followed by MBcoPA4 (7.07×10^{-5} S/m), MBcoPA1 (2.95×10^{-4} S/m) and with the highest for MBcoPA3 (4.87×10^{-3} S/m). These results are consistent with the macrodispersion of MWCNTs presented in Table 2 and DC electrical conductivity results shown in Figure 3. Conductivity spectra obey the Almond–West power law:

$$\sigma(\omega) = \sigma_{DC} + \left(\frac{\omega}{\omega_{cr}}\right)^s \quad (1)$$

where σ_{DC} is the DC conductivity, ω_{cr} is the critical frequency at which the conductivity $\sigma(\omega)$ from the DC plateau, and s is the parameter. From Figure 4a, it is possible to conclude that the behavior of critical frequency is correlated with the behavior of the DC conductivity, with no correlation with behavior of parameter s. A factor which is important for the frequency-dependent conductivity $\sigma(\omega)$ is the electron transport not only across the whole sample but also inside some conductive clusters, if the electron flight time τ inside the cluster is smaller than the reciprocal electromagnetic wave frequency $1/2\pi\omega$. Thus, bigger aggregates are related with smaller critical frequencies, while parameter s is related with the distribution of aggregates' size [52] (Figure 4a and Table 2).

The incorporation of MWCNTs led to an increase in the dielectric permittivity, by about two orders of magnitude for masterbatches based on coPA1, coPA2 and coPA3, and much more (10^6) for coPA3. The behavior of dielectric permittivity at low frequencies (for example 129 Hz) is correlated with the DC conductivity behavior, except in sample MBcoPA2. The dielectric permittivity of this sample is higher than the dielectric permittivity of samples MBcoPA1 and MbcoPA4. Such mismatch in the behavior of dielectric permittivity and electrical conductivity was already explained by the difference in distributions of relaxation times and distributions of agglomerate size [52]. Indeed, the distribution of agglomerate size is much broader for sample MBcoPA2 than for samples MBcoPA1 and MBcoPA4 (see Table 2). Together with the frequency, the dielectric permittivity decreased as was reported for thermoplastics [53], elastomers [54] and thermoplastic elastomers [55] containing CNTs.

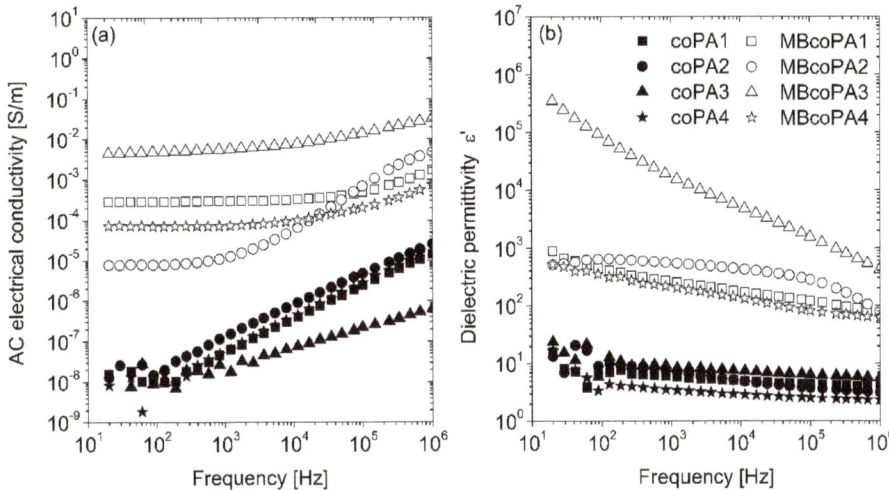

Figure 4. Comparison of: (**a**) the real part of the electrical conductivity; and (**b**) the real part of the dielectric permittivity for the neat coPAs and their masterbatches containing 7 wt % MWCNT.

It has been observed previously using the TGA method, that in the presence of CNTs, the flux of degradation products is hindered and delays the start of the degradation process of the polymer [56]. The influence of the addition of MWCNTs on the coPAs' thermal stability was examined by TGA and the results are presented in Table 3. The example TGA curves for coPA and the masterbatch are presented in Figure 5. It is seen that for all pure coPAs at 2% weight loss, the decomposition starts below 200 °C with the lowest temperature $T_{2\%}$ for coPA3. In the presence of 7 wt % MWCNT, the decomposition temperature ($T_{2\%}$) rises the most for MBcoPA1 and for MBcoPA2 by about 103 °C and 88 °C, respectively. Interestingly, for MBcoPA3 and MBcoPA4, the increase is only a few degrees. For them, higher increase in the decomposition temperatures was determined at 5 wt % weight loss, about 23 °C (MBcoPA3) and 15 °C (MBcoPA4). It is worth noting that the decomposition process in pure coPAs starts at a much lower temperature than that reported for the typical PA6 and PA66 homopolymers which decompose at 350 °C [57]. The maximum rates of weight loss for pure coPAs occurs at 450 °C and this temperature raises slightly (6–17 °C) in the presence of 7 wt % MWCNT. It seems that such a high amount of MWCNTs has a smaller effect on the thermal stability of the polymer than the lower concentration. The PA6 composites mixed with only 0.5 wt % MWCNT caused the shift of about 70 °C in the temperature decomposition [58].

The inclusion of CNTs in the polymer alters the nucleation process, resulting in formation of a crystalline phase which in turn can affect the electrical properties of the nanocomposites. The DSC method was used to determine the thermal properties of the materials listed in Table 4 and collected in Figure 6. On the post-processing, the first heating curves (Figure 6a), two peaks from the glass transition (T_g) and the melting point (T_m) were detected. For the unfilled coPA1 and coPA3, T_g occurs at a lower temperature of around 50 °C, while for coPA2 and coPA4, at a higher temperature of around 70 °C. Because selected coPAs consist of the PA6/PA66 copolyamide, the lower T_g is probably associated with PA6 segments; in turn, the higher T_g temperature comes from PA66 segments [59]. The addition of 7 wt % MWCNT shifts the T_g peak towards higher values since well dispersed MWCNTs hamper the mobility of the polymer chains [60]. Melting points for pure coPAs are much lower than for typical PA6 and PA66 homopolymers, between 110 and 130 °C, as visible during the first and second heating curves (Figure 6a,b). Due to the complex formulation of coPAs, the melting peak is broad, especially for coPA3, meaning that the crystal phases are not homogenous. Moreover, for all coPAs except coPA3, there is clear evidence for the cold crystallization visible as an exothermic peak and it can be associated with the too slow crystallization process. After the addition of 7 wt % MWCNT, the melting peak is shifted but only by a few degrees (2–6 °C) meaning that their application temperature (180 °C according to Materials Safety Data Sheet) will remain unchanged. The nucleating role of MWCNTs is also confirmed by the cooling curves presented in Figure 6c. Interestingly, pure coPAs do not possess any thermal processes. However, for 7 wt % MWCNT masterbatches, broad peaks appear at the maximum of around 85 °C for MBcoPA1, MBcoPA1 and MBcoPA3 and with a sharper peak for MBcoPA4 with the maximum at 92 °C. The sharper the peak, the more perfect crystals are formed which come from one type of polyamide segment [61]. In comparison to typical homopolymers in which the crystallization is more clear due to their homogenous structure, in HMAs, the crystallization process may be disturbed by the presence of various components [59–61]. The formation of the new crystal phase can be noticed by the changes in the enthalpy of melting, ΔH_m. There is the same decreasing in the enthalpy of melting determined at the first and second heating after the incorporation of 7 wt % MWCNT. At the first heating, the most for the MBcoPA1 was about 31.8 J/g, then for coPA1 about 14.7 J/g and for coPA2 and coPA4, it was 10.4 J/g and 8 J/g, respectively. During the second heating, these differences were smaller, with only a few J/g except for MBcoPA2 where the ΔHm decreased by about 21.4 J/g.

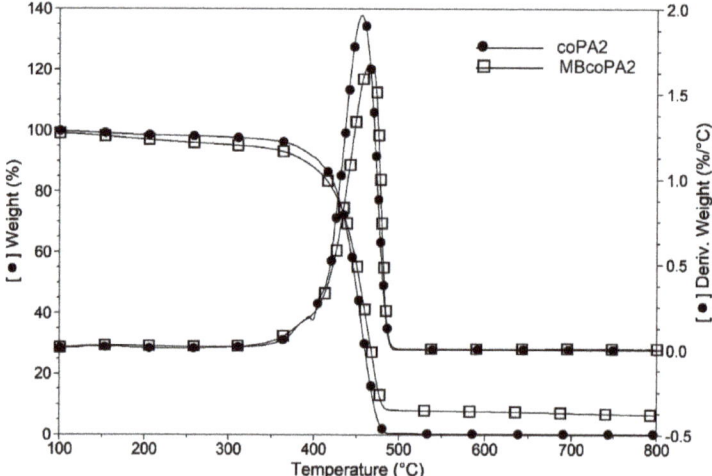

Figure 5. The example TGA curve for the neat coPA2 and its masterbatch containing 7 wt % MWCNT.

Table 4. Summary of the thermal analysis results. "—" means a lack of peak on the curve.

Material	TGA			DSC					
				First Heating			Second Heating		Cooling
	$T_{2\%}$ (°C)	$T_{5\%}$ (°C)	T_d (°C)	T_g (°C)	T_m (°C)	ΔH_m (J/g)	T_m (°C)	ΔH_m (J/g)	T_c (°C)
coPA1	188	339	455	46.1	130	64.1	128	31.8	—
MBcoPA1	291	379	461	50.1	133	32.3	133	29.0	92.7
coPA2	184	337	455	70.6	121	51.9	124	40.4	—
MBcoPA2	272	376	464	85.5	124	37.2	126	19.0	86.4
coPA3	171	271	443	52.8	111	25.1	110	16.8	—
MBcoPA3	178	294	457	70.5	116	14.7	116	15.3	84.8
coPA4	196	346	447	72.5	120	31.4	121	25.9	—
MBcoPA4	201	361	464	72.7	122	23.4	122	25.8	87.1

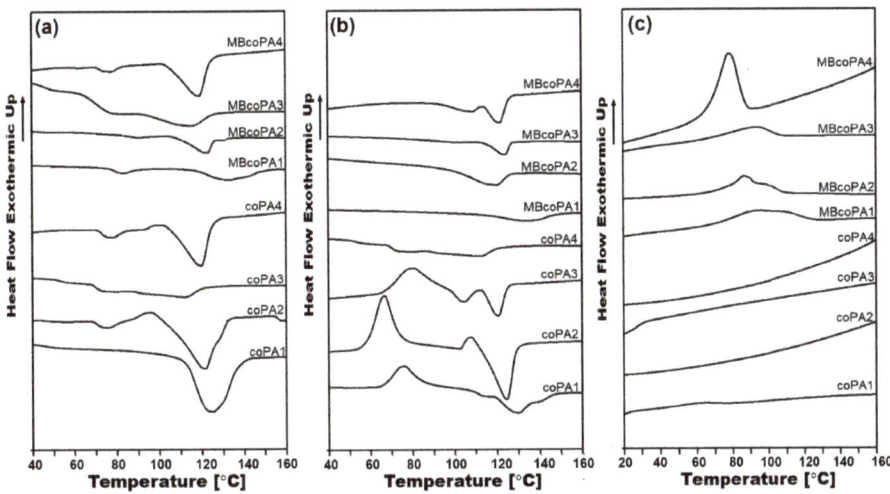

Figure 6. (a) First heating curves; (b) second heating curves; and (c) cooling curves for the unfilled coPAs and their masterbatches containing 7 wt % MWCNT.

In order to see the differences in the adhesion properties of the pure coPAs and their masterbatches with 7 wt % MWCNT, the analysis of the contact angle and surface energy was performed. According to the results collected in Table 5, among all the studied coPAs, coPA3 had a significantly higher contact angle and lower surface energy than the other coPAs. The differences in the wettability of the selected coPAs were related to their specific compositions; especially the type and content of the tackifier [62]. Because CNTs change the surface properties of the polymers, the wettability of the nanocomposites will be not the same as of the pure coPAs. As shown in Table 4, the inclusion of 7 wt % MWCNT resulted in the greatest decrease in the contact angle for MBcoPA2 by about 31°, followed by a decrease of 19° for MBcoPA3 and the lowest for MBcoPA1 by about 7°. The calculated surface energy for these masterbatches increased in the same order as an effect of the modification of the surface by the addition of MWCNTs. Yang at al. also reported an improvement of the hydrophilicity of the gutta-percha nanocomposites at 2 wt % MWCNT content [63]. It should be noted that the values of the contact angle in coPA1 and coPA2 as well as their masterbatches indicate the hydrophilic character of their surfaces, resulting in good adhesion and wettability. For coPA3, the contact angle was 99°, which means that this coPA was more hydrophobic, but in the presence of MWCNTs, the surface became hydrophilic since the contact angle decreased to 80°. This was the opposite to coPA4, in which the unfilled polymer was more hydrophilic but the addition of MWCNTs promoted the hydrophobic surface properties (contact angle >90°) associated with the poor adhesion and wettability.

Table 5. Variation in the contact angle and surface energy for the pure coPAs and their masterbatches.

Material	Average Contact Angle (°)	Average Surface Energy (mN/m)
coPA1	85 ± 1.5	34.02 ± 0.005
MBcoPA1	78 ± 0.6	36.61 ± 0.004
coPA2	83 ± 5.0	33.44 ± 0.002
MBcoPA2	52 ± 3.0	52.81 ± 0.004
coPA3	99 ± 4.2	23.72 ± 0.005
MBcoPA3	80 ± 3.5	35.45 ± 0.006
coPA4	88 ± 3.1	30.43 ± 0.003
MBcoPA4	96 ± 0.6	25.92 ± 0.004

4. Conclusions

The present paper describes a new group of electrically conducive adhesives fabricated by the melt-blending of thermoplastic hot melt adhesives based on coPAs and 7 wt % MWCNT. The selected coPAs have low, medium and high viscosity which together with the loss and storage modulus was increased by about 3–6 orders of magnitude as an effect of the strong interactions occurring between coPA macromolecules and MWCNTs. It was confirmed that viscosity has no effect the on the dispersion and distribution of MWCNTs in the polymer matrix. From microscopic images, the area ratio of the agglomerates calculated by ImageJ was the lowest for coPA3 having MVR = 350 (A_A = 4.18%) and the highest for coPA2 (A_A = 11.8%) and coPA4 (A_A = 11.6%) having MVR = 600 and 150, respectively. The correlation of the determined number of agglomerates with the measured electrical conductivity of the coPAs + 7 wt % MWCNT clearly presented an increase in the electrical conductivity value when A_A decreased. Therefore, the highest DC electrical conductivity was achieved for coPA3 + 7 wt % MWCNT; σ = 0.67 S/m and the lowest for coPA2 + 7 wt % MWCNT σ = 0.0076 S/m. These results are consistent with the AC electrical conductivity analyzed by dielectric spectroscopy which also showed an increase in the dielectric permittivity in the presence of MWCNTs. The addition of 7 wt % MWCNT to coPAs shifted the decomposition temperature towards higher values, especially for coPA1 and coPA2, to about 103 °C and 88 °C, respectively. However, the thermal stability increased by only a few degrees since high MWCNT content is not as effective as a low concentration. Similarly, the melting points of coPAs + 7 wt % MWCNT increased by a few degrees (2–6 °C) which, together with the changes in the enthalpy of melting, indicates the nucleation effect of MWCNTs. The adhesion properties analyzed by the measurement of the contact angle and the surface energy indicated that depending on the type of coPA used and its composition, they are more hydrophobic or hydrophilic. The addition of MWCNTs modifies the surface of the nanocomposites visible by the enhancement of hydrophobicity or hydrophilicity of the coPAs. Thermoplastic hot melt copolyamides containing MWCNTs are the example of the ECAs which can be used to join the composite structures together to provide the conductive interlayer required for lightning strike protection.

Author Contributions: Conceptualization, P.L.-D.; methodology, P.L.-D, R.K., J.M. and K.D.; software, P.L.-D.; validation, P.L.-D and A.B.; formal analysis, P.L.-D and A.B.; investigation, P.-L.-D., R.K., J.M. and K.D.; resources, P.L.-D.; data curation, P.L-D.; writing—original draft preparation, P.L.-D.; writing—review and editing, J.M. and A.B.; visualization, P.L.-D.; supervision, P.L.-D.; project administration, A.B.; funding acquisition, A.B. All authors have read and agreed to the published version of the manuscript.

Funding: This research was funded by EUROPEAN UNION within the project entitled Open Access Pilot Plants for Sustainable Industrial Scale Nanocomposites Manufacturing Based on Buckypapers, Doped Veils and Prepregs; grant number 646307.

Acknowledgments: The authors would like to thank Nadir Kchnit from Nanocyl, Belgium for the fabrication of masterbatches pellets.

Conflicts of Interest: The authors declare no conflict of interest.

References

1. Nele, L.; Palmieri, B. Electromagnetic heating for adhesive melting in CFRTP joining: Study, analysis, and testing. *Int. J. Adv. Manuf. Technol.* **2020**, *106*, 5317–5331. [CrossRef]
2. Peng, X.; Liu, S.; Huang, Y.; Sang, L. Investigation of joining of continuous glass fibre reinforced polypropylene laminates via fusion bonding and hotmelt adhesive film. *Int. J. Adhes. Adhes.* **2020**, *100*, 102615. [CrossRef]
3. Ciardiello, R.; Belingardi, G.; Martorana, B.; Brunella, V. Physical and mechanical properties of a reversible adhesive for automotive applications. *Int. J. Adhes. Adhes.* **2019**, *89*, 117–128. [CrossRef]
4. Ganesh, M.G.; Lavenya, K.; Kirubashini, K.A.; Ajeesh, G.; Bhowmik, S.; Epaarachchi, J.A.; Yuan, X. Electrically conductive nano adhesive bonding: Futuristic approach for satellites and electromagnetic interference shielding. *Adv. Aircr. Spacecr. Sci.* **2017**, *4*, 729–744. [CrossRef]
5. Geipel, T.; Meinert, M.; Kraft, A.; Eitner, U. Optimization of electrically conductive adhesive bonds in photovoltaic modules. *IEEE J. Photovolt.* **2018**, *8*, 1074–1081. [CrossRef]

6. Lopes, P.E.; Moura, D.; Freitas, D.; Proença, M.F.; Figueiredo, H.; Alves, R.; Paiva, M.C. Advanced electrically conductive adhesives for high complexity PCB assembly. *AIP Conf. Proc.* **2019**, *2055*. [CrossRef]
7. Yim, M.J.; Li, Y.; Moon, K.S.; Paik, K.W.; Wong, C.P. Review of recent advances in electrically conductive adhesive materials and technologies in electronic packaging. *J. Adhes. Sci. Technol.* **2008**, *22*, 1593–1630. [CrossRef]
8. Dong, H.; Li, X.; Dong, Y.; Guo, S.; Zhao, L. A novel preparation method of electrically conductive adhesives by powder spraying process. *Materials* **2019**, *12*, 2793. [CrossRef]
9. Hao, J.; Wang, D.; Li, S.; He, X.; Zhou, J.; Xue, F. The effect of conductive filler on the properties of Electrically Conductive Adhesives (ECAs). In Proceedings of the 18th International Conference on Electronic Packaging Technology, Harbin, China, 16–19 August 2017; pp. 803–808. [CrossRef]
10. Lu, J.; Liu, D.; Dai, J. Preparation of highly conductive silver nanowires for electrically conductive adhesives. *J. Mater. Sci. Mater. Electron.* **2019**, *30*, 15786–15794. [CrossRef]
11. Ma, H.; Li, Z.; Tian, X.; Yan, S.; Li, Z.; Guo, X.; Ma, Y.; Ma, L. Silver flakes and silver dendrites for hybrid electrically conductive adhesives with enhanced conductivity. *J. Electron. Mater.* **2018**, *47*, 2929–2939. [CrossRef]
12. Jing, L.; Lumpp, J.K. Electrical and mechanical characterization of carbon nanotube filled conductive adhesive. In Proceedings of the 2006 IEEE Aerospace Conference, Big Sky, MT, USA, 4–11 March 2006; p. 6.
13. Pu, N.W.; Peng, Y.Y.; Wang, P.C.; Chen, C.Y.; Shi, J.N.; Liu, Y.M.; Ger, M.D.; Chang, C.L. Application of nitrogen-doped graphene nanosheets in electrically conductive adhesives. *Carbon* **2014**, *67*, 449–456. [CrossRef]
14. Logakis, E.; Pandis, C.; Peoglos, V.; Pissis, P.; Pionteck, J.; Pötschke, P.; Mičušík, M.; Omastová, M. Electrical/dielectric properties and conduction mechanism in melt processed polyamide/multi-walled carbon nanotubes composites. *Polymer* **2009**, *50*, 5103–5111. [CrossRef]
15. Strozzi, M.; Pellicano, F. Linear vibrations of triple-walled carbon nanotubes. *Math. Mech. Solids* **2018**, *23*, 1456–1481. [CrossRef]
16. Dai, H.; Wong, E.W.; Liebert, C.M.; Lieber, C.M. Probing electrical transport in nanomaterials: Conductivity of individual carbon nanotubes. *Science* **1996**, *272*, 523–526. [CrossRef]
17. Li, J.; Lumpp, J.K.; Andrews, R.; Jacques, D. Aspect ratio and loading effects of multiwall carbon nanotubes in epoxy for electrically conductive adhesives. *J. Adhes. Sci. Technol.* **2008**, *22*, 1659–1671. [CrossRef]
18. Aradhana, R.; Mohanty, S.; Nayak, S.K. High performance electrically conductive epoxy/reduced graphene oxide adhesives for electronics packaging applications. *J. Mater. Sci. Mater. Electron.* **2019**, *30*, 4296–4309. [CrossRef]
19. Meng, Q.; Han, S.; Araby, S.; Zhao, Y.; Liu, Z.; Lu, S. Mechanically robust, electrically and thermally conductive graphene-based epoxy adhesives. *J. Adhes. Sci. Technol.* **2019**, *33*, 1337–1356. [CrossRef]
20. McClory, C.; McNally, T.; Baxendale, M.; Pötschke, P.; Blau, W.; Ruether, M. Electrical and rheological percolation of PMMA/MWCNT nanocomposites as a function of CNT geometry and functionality. *Eur. Polym. J.* **2010**, *135*, 854–868. [CrossRef]
21. Socher, R.; Krause, B.; Müller, M.T.; Boldt, R.; Pötschke, P. The influence of matrix viscosity on MWCNT dispersion and electrical properties in different thermoplastic nanocomposites. *Polymer* **2012**, *53*, 495–504. [CrossRef]
22. Sumita, M.; Abe, H.; Kayaki, H.; Miyasaka, K. Effect of melt viscosity and surface tension of polymers on the percolation threshold of conductive-particle-filled polymeric composites. *J. Macromol. Sci.* **1986**, *25*, 171–184. [CrossRef]
23. Zhang, Y.; Zhang, F.; Xie, Q.; Wu, G. Research on electrically conductive acrylate resin filled with silver nanoparticles plating multiwalled carbon nanotubes. *J. Reinf. Plast. Compos.* **2015**, *34*, 1193–1201. [CrossRef]
24. Dang, Z.-M.; Shehzad, K.; Zha, J.-W.; Mujahid, A.; Hussain, T.; Nie, J.; Shi, C.-Y. Complementary percolation characteristics of carbon fillers based electrically percolative thermoplastic elastomer composites. *Compos. Sci. Technol.* **2011**, *72*, 28–35. [CrossRef]
25. Marcq, F.; Demont, P.; Monfraix, P.; Peigney, A.; Laurent, C.; Falat, T.; Courtade, F.; Jamin, T. Carbon nanotubes and silver flakes filled epoxy resin for new hybrid conductive adhesives. *Microelectron. Reliab.* **2011**, *51*, 1230–1234. [CrossRef]

26. Cui, H.W.; Li, D.S.; Fan, Q. Using a functional epoxy, micron silver flakes, nano silver spheres, and treated single-wall carbon nanotubes to prepare high performance electrically conductive adhesives. *Electron. Mater. Lett.* **2013**, *9*, 299–307. [CrossRef]
27. Ma, H.; Qiu, H.; Qi, S. Electrically conductive adhesives based on acrylate resin filled with silver-plated graphite nanosheets and carbon nanotubes. *J. Adhes. Sci. Technol.* **2015**, *29*, 2233–2244. [CrossRef]
28. Troughton, M. Adhesive bonding. In *Handbook of Plastics Joining*; William Andrew Inc.: Norwich, NY, USA, 2008; pp. 145–173.
29. Nobile, M.R. Rheology of polymer–carbon nanotube composites melts. In *Polymer–Carbon Nanotube Composites Preparation, Properties and Applications*; Woodhead Publishing Limited: Sawston, UK, 2011; pp. 428–481. ISBN 978-1-84569-761-7.
30. Brewis, D. Hot melt adhesives. In *Handbook of Adhesion*; Wiley: Hoboken, NJ, USA, 2005; pp. 711–757. ISBN 9780470014226.
31. Ebnesajjad, S. Characteristics of adhesive materials. In *Handbook of Adhesives and Surface Preparation*; Ebnesajjad, S., Ed.; Elsevier Inc.: Amsterdam, The Netherlands, 2011; pp. 137–183. ISBN 9781437744613.
32. Latko-Durałek, P.; McNally, T.; Macutkevic, J.; Kay, C.; Boczkowska, A. Hot-melt adhesives based on co-polyamide and multiwalled carbon nanotubes. *J. Appl. Polym. Sci.* **2017**, *1*, 1–15. [CrossRef]
33. Paul, C.W. Hot Melt Adhesives for Dermal Application. U.S. Patent 6,448,303, 10 September 2002.
34. Kalish, J.P.; Ramalingam, S.; Bao, H.; Hall, D.; Wamuo, O.; Ling, S.; Paul, C.W.; Eodice, A. An analysis of the role of wax in hot melt adhesives. *Int. J. Adhes. Adhes.* **2015**, *60*, 63–68. [CrossRef]
35. Fernández, M.; Landa, M.; Muñoz, M.E.; Santamaría, A. Tackiness of an electrically conducting polyurethanenanotube nanocomposite. *Int. J. Adhes. Adhes.* **2010**, *30*, 609–614. [CrossRef]
36. Fernández, M.; Landa, M.; Muñoz, M.E.; Santamaría, A. Thermal and viscoelastic features of new nanocomposites based on a hot-melt adhesive polyurethane and multi-walled carbon nanotubes. *Macromol. Mater. Eng.* **2010**, *295*, 1031–1041. [CrossRef]
37. Landa, M.; Canales, J.; Fernández, M.; Muñoz, M.E.; Santamaría, A. Effect of MWCNTs and graphene on the crystallization of polyurethane based nanocomposites, analyzed via calorimetry, rheology and AFM microscopy. *Polym. Test.* **2014**, *35*, 101–108. [CrossRef]
38. Landa, M.; Fernández, M.; Muñoz, M.E.; Santamaría, A. The effect of flow on the physical properties of polyurethane/carbon nanotubes nanocomposites:repercussion on their use as electrically conductive hot-melt adhesives. *Polym. Compos.* **2015**, 704–712. [CrossRef]
39. Canales, J.; Muñoz, M.E.; Fernández, M.; Santamaría, A. Rheology, electrical conductivity and crystallinity of a polyurethane/graphene composite: Implications for its use as a hot-melt adhesive. *Compos. Part. A Appl. Sci. Manuf.* **2016**, *84*, 9–16. [CrossRef]
40. Wehnert, F.; Pötschke, P.; Jansen, I. Hotmelts with improved properties by integration of carbon nanotubes. *Int. J. Adhes. Adhes.* **2015**, *62*, 63–68. [CrossRef]
41. Cecen, V.; Boudenne, A.; Ibos, L.; Novák, I.; Nógellová, Z.; Prokeš, J.; Krupa, I. Electrical, mechanical and adhesive properties of ethylene-vinylacetate copolymer (EVA) filled with wollastonite fibers coated by silver. *Eur. Polym. J.* **2008**, *44*, 3827–3834. [CrossRef]
42. Pomposo, J.A.; Rodríguez, J.; Grande, H. Polypyrrole-based conducting hot melt adhesives for EMI shielding applications. *Synth. Met.* **1999**, *104*, 107–111. [CrossRef]
43. Vlachopoulos, J.; Strutt, D. The role of rheology in polymer extrusion. *New Technol. Extrus.* **2003**, 1–26. [CrossRef]
44. Vlachopoulos, J.; Strutt, D. Rheology of molten polymers. In *Multilayer Flexible Packaging*; Elsevier Inc.: Amsterdam, The Netherlands, 2010; pp. 57–72. ISBN 9780815520214.
45. Zabegaeva, O.N.; Sapozhnikov, D.A.; Buzin, M.I.; Krestinin, A.V.; Kotelnikov, V.A.; Baiminov, B.A.; Afanasyev, E.S.; Pashunin, Y.M.; Vygodskii, Y.S. Nylon-6 and single-walled carbon nanotubes polyamide composites. *High Perform. Polym.* **2017**, *29*, 411–421. [CrossRef]
46. Ha, H.; Kim, S.C.; Ha, K. Morphology and properties of polyamide/multi-walled carbon nanotube composites. *Macromol. Res.* **2010**, *18*, 660–667. [CrossRef]
47. Li, Y.; Shimizu, H. High-shear melt processing of polymer–carbon nanotube composites. *Polym. Nanotub. Compos. Prep. Prop. Appl.* **2011**, 133–154. [CrossRef]
48. Kasaliwal, G.R.; Göldel, A.; Pötschke, P.; Heinrich, G. Influences of polymer matrix melt viscosity and molecular weight on MWCNT agglomerate dispersion. *Polymer* **2011**, *52*, 1027–1036. [CrossRef]

49. Kim, S.; Drzal, L.T. Comparison of exfoliated graphite nanoplatelets (xGnP) and CNTs for reinforcement of EVA nanocomposites fabricated by solution compounding method and three screw rotating systems. *J. Adhes. Sci. Technol.* **2009**, *23*, 1623–1638. [CrossRef]
50. Beaume, F.; Lauprêtre, F.; Monnerie, L.; Maxwell, A.; Davies, G.R. Secondary transitions of aryl-aliphatic polyamides. I. Broadband dielectric investigation. *Polymer* **2000**, *41*, 2677–2690. [CrossRef]
51. Curtis, A.J. Dielectric Properties of Polyamides: Polyhexamethylene Adiparnide and Polyhexarnethylene Sebacarnide. *J. Res. Nat. Bur. Stand. Sect. A Phys. Chem.* **1961**, *65*, 185. [CrossRef] [PubMed]
52. Bertasius, P.; Meisak, D.; Macutkevic, J.; Kuzhir, P.; Selskis, A.; Volnyanko, E.; Banys, J. Fine tuning of electrical transport and dielectric properties of epoxy/carbon nanotubes composites via magnesium oxide additives. *Polymers* **2019**, *11*, 2044. [CrossRef]
53. Yadav, P.; Srivastava, A.K.; Yadav, M.K.; Kripal, R.; Singh, V.; Lee, D.B.; Lee, J.H. Synthesis and dielectric characterization of polycarbonate/multi-wall carbon nanotubes nanocomposite. *Arab. J. Chem.* **2015**, *12*, 440–446. [CrossRef]
54. Vilčáková, J.; Moučka, R.; Svoboda, P.; Ilčíková, M.; Kazantseva, N.; Hřibová, M.; Mičušík, M.; Omastová, M. Effect of surfactants and manufacturing methods on the electrical and thermal conductivity of carbon nanotube/silicone composites. *Molecules* **2012**, *17*, 13157–13174. [CrossRef]
55. Meier, J.G.; Crespo, C.; Pelegay, J.L.; Castell, P.; Sainz, R.; Maser, W.K.; Benito, A.M. Processing dependency of percolation threshold of MWCNTs in a thermoplastic elastomeric block copolymer. *Polymer* **2011**, *52*, 1788–1796. [CrossRef]
56. Liu, T.X.; Huang, S. 15-Morphology and thermal behavior of polymer/carbon nanotube composites. *Woodhead Publ. Ser. Compos. Sci. Eng.* **2010**, 529–562. [CrossRef]
57. Jang, B.N.; Wilkie, C.A. The effect of clay on the thermal degradation of polyamide 6 in polyamide 6/clay nanocomposites. *Polymer* **2005**, *46*, 3264–3274. [CrossRef]
58. Mahmood, N.; Islam, M.; Hameed, A.; Saeed, S. Polyamide 6/multiwalled carbon nanotubes nanocomposites with modified morphology and thermal properties. *Polymers* **2013**, *5*, 1380–1391. [CrossRef]
59. Greco, R.; Nicolais, L. Glass transition temperature in nylons. *Polymer* **1976**, *17*, 1049–1053. [CrossRef]
60. Xie, X.L.; Mai, Y.W.; Zhou, X.P. Dispersion and alignment of carbon nanotubes in polymer matrix: A review. *Mater. Sci. Eng. R Rep.* **2005**, *49*, 89–112. [CrossRef]
61. Brosse, A.-C.; Tencé-Girault, S.; Piccione, P.M.; Leibler, L. Effect of multi-walled carbon nanotubes on the lamellae morphology of polyamide-6. *Polymer* **2008**, *49*, 4680–4686. [CrossRef]
62. Fernandes, E.G.; Lombardi, A.; Solaro, R.; Chiellini, E. Thermal characterization of three-component blends for hot-melt adhesives. *J. Appl. Polym. Sci.* **2001**, *80*, 2889–2901. [CrossRef]
63. Yang, H.; Peng, P.; Sun, Q.; Zhang, Q.; Ren, N.; Han, F.; She, D. Developed carbon nanotubes/gutta percha nanocomposite films with high stretchability and photo-thermal conversion efficiency. *J. Mater. Res. Technol.* **2020**, *9*, 8884–8895. [CrossRef]

© 2020 by the authors. Licensee MDPI, Basel, Switzerland. This article is an open access article distributed under the terms and conditions of the Creative Commons Attribution (CC BY) license (http://creativecommons.org/licenses/by/4.0/).

Article

Mechanical and Electrical Properties of Epoxy Composites Modified by Functionalized Multiwalled Carbon Nanotubes

Paweł Smoleń [1,2,*], Tomasz Czujko [1,*], Zenon Komorek [1], Dominik Grochala [3], Anna Rutkowska [2] and Małgorzata Osiewicz-Powęzka [4]

Citation: Smoleń, P.; Czujko, T.; Komorek, Z.; Grochala, D.; Rutkowska, A.; Osiewicz-Powęzka, M. Mechanical and Electrical Properties of Epoxy Composites Modified by Functionalized Multiwalled Carbon Nanotubes. *Materials* 2021, 14, 3325. https://doi.org/10.3390/ma14123325

Academic Editor: Werner Blau

Received: 18 May 2021
Accepted: 14 June 2021
Published: 16 June 2021

Publisher's Note: MDPI stays neutral with regard to jurisdictional claims in published maps and institutional affiliations.

Copyright: © 2021 by the authors. Licensee MDPI, Basel, Switzerland. This article is an open access article distributed under the terms and conditions of the Creative Commons Attribution (CC BY) license (https://creativecommons.org/licenses/by/4.0/).

[1] Institute of Materials Science and Engineering, Military University of Technology, Gen. Sylwestra Kaliskiego Str. 2, 00-908 Warsaw, Poland; zenon.komorek@wat.edu.pl
[2] Smart Nanotechnologies S.A., Karola Olszewskiego Str. 25, 32-566 Alwernia, Poland; anna.rutkowska@smartnanotech.com.pl
[3] Department of Electronics, AGH University of Science and Technology, Adama Mickiewicza Str. 30, 30-059 Cracow, Poland; grochala@agh.edu.pl
[4] New Era Materials Sp. z o.o., Komandosów Str. 1/7, 32-085 Modlniczka, Poland; malgorzata.osiewicz@neweramaterials.com
* Correspondence: pawel.smolen@wat.edu.pl or pawel.smolen@smartnanotech.com.pl (P.S.); tomasz.czujko@wat.edu.pl (T.C.); Tel.: +48-783-123-416 (P.S.); +48-261-839-445 (T.C.)

Abstract: This paper investigates the effect of multiwalled carbon nanotubes on the mechanical and electrical properties of epoxy resins and epoxy composites. The research concerns multiwalled carbon nanotubes obtained by catalytic chemical vapor deposition, subjected to purification processes and covalent functionalization by depositing functional groups on their surfaces. The study included the analysis of the change in DC resistivity, tensile strength, strain, and Young's modulus with the addition of carbon nanotubes in the range of 0 to 2.5 wt.%. The effect of agents intended to increase the affinity of the nanomaterial to the polymer on the aforementioned properties was also investigated. The addition of functionalized multiwalled carbon nanotubes allowed us to obtain electrically conductive materials. For all materials, the percolation threshold was obtained with 1% addition of multiwalled carbon nanotubes, and filling the polymer with a higher content of carbon nanotubes increased its conductivity. The use of carbon nanotubes as polymer reinforcement allows higher values of tensile strength and a higher strain percentage to be achieved. In contrast, Young's modulus values did not increase significantly, and higher nanofiller percentages resulted in a drastic decrease in the values of the abovementioned properties.

Keywords: polymer composites; multiwall carbon nanotubes; electrical properties; mechanical properties; functionalization; epoxy

1. Introduction

Polymer nanocomposites are materials that consist of two or more phases (continuous and discontinuous) with a distinct interaction surface, of which at least one component has at least one dimension that is nanometric in scale. In contrast to conventional polymer composites based on micron-scale modifiers, the introduction of nanomaterials into polymers allows for small filler distances, so that composite properties can be modified to a large extent even at very low additive contents [1,2]. The nanofillers used can be divided according to their chemical properties (inorganic and organic), physical structure (crystalline, amorphous, and gas inclusions), and particle shape (three-, two- and one-dimensional). Depending on the type of polymer and nanofiller, polymer nanocomposites have applications as structural, functional, and coating materials [3]. The key aspect related to the effective use of nanomaterials in polymeric materials is their homogeneous dispersion in the matrix while maintaining strong interaction and adhesion between the polymer chain and nanofiller. The size of the contact surface of the discontinuous phase (filler), as well as the nature of the interactions between the continuous and discontinuous phases, plays a vital role [4,5].

Carbon nanotubes are one-dimensional materials characterized by a high length-to-diameter ratio that can be as high as 1000. They are made of rolled single sheets of graphite, referred to as graphene layers, that are a single layer of graphite-structured carbon atoms forming six-membered rings of carbon atoms with sp^2 hybridization [6–8]. Their unique structure also gives unique properties, e.g., good mechanical and electrical properties, and taking into account other properties, such as low density, high expansion coefficient, or large specific surface area, they might be considered interesting reinforcing materials for composite materials [4,7].

An improvement in the mechanical and electrical properties is obtained when the length-to-diameter ratio of the filler increases, and when its transverse dimension decreases. In this case, the specific surface area of the filler increases, and thus, the interaction between the matrix and its particles increases [3]. Adhesion between the nanofiller and the polymer can be improved by functionalization of carbon nanotubes [9,10]. In the case of carbon nanotubes, the most common methods for the surface modification are physicochemical methods, which include covalent and noncovalent interactions [11].

Grafting of carboxyl groups, fluorinated nanotubes and nanotubes functionalized with isocyanates, amino groups, or biomolecules is often used to improve interactions and mechanical properties [6]. A number of review papers presenting the current knowledge on the mechanical properties of polymers with carbon nanotubes have been written [5,12]. The effect of the addition of carbon nanotubes on the mechanical properties has been discussed in detail in the work of Coleman et al. [13], who analyzed the influence of the composite processing method on the strength parameters of the composites obtained, while the work of Domun et al. [11] focused on the issue of improving the fracture toughness and strength of epoxy resin using nanomaterials, particularly carbon nanotubes.

Gojny [14] carried out an extensive comparative study that involved the changes in mechanical properties through the use of single-, double-, and multiwalled carbon nanotubes modified with amino groups, and examined the effect of adding carbon nanotubes to an epoxy resin. The use of double-walled carbon nanotubes with amino groups resulted in an improvement in tensile strength by 10% and Young's modulus by 15%. On the other hand, Spitalsky's research [15] proved that the method of dispersing 0.5% by weight of multiwalled carbon nanotubes in a solvent improved the tensile strength by 62% with an improvement in Young's modulus by 54%. The effect of increasing the strength of epoxy composites has also been attempted using oxidized carbon nanotubes. In the work of Gou et al., a 40% increase in tensile strength and a 10% decrease in Young's modulus values were reported using 4 wt.% carbon nanotubes introduced into the matrix using the stir method [16]. An opposite effect was obtained by Breton [17]—at 6% addition, a 32% increase in Young's modulus and a 25% decrease in tensile strength were obtained.

The potential of carbon nanotubes is also often used in studies of the electrical properties of polymer nanocomposites. Carbon nanotubes, due to their specific structure and excellent electrical conductivity, make it possible to obtain a significant improvement in the electrical conductivity of polymers at a very low-weight fraction. A careful analysis of the literature indicates a large variation in the values of the electrical properties, as the electrical conductivity and the percolation threshold are highly dependent on many factors, including the type of polymer, the method of nanocomposite synthesis, the length-to-diameter ratio of the carbon nanotube, the degree of dispersion, or the functionalization of the additive [5,18]. Figure 1 presents conductivity ranges of conducting polymers and also conductive materials.

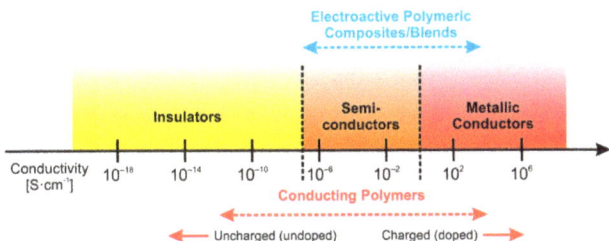

Figure 1. Conductivity range of conducting polymers and conductive materials based on [19].

Establishing epoxy composites as the focal point of their study, Sandler et al. presented an improvement in electrical conductivity with 0.0025% unmodified carbon nanotubes with a length-to-diameter ratio of 340 added by hot shearing [20]. On the other hand, in the work of Kovacs [21], the percolation threshold was reached at 0.011% and 0.08% of unmodified carbon nanotubes using slow and fast shear mixing, respectively, proving that a slower mixing speed significantly increased the conductivity at the same length-to-diameter ratio of carbon nanotubes.

On the other hand, carbon nanotube dispersions prepared by sonication and mixing in epoxy resin resulted in obtaining carbon nanotubes with a low length-to-diameter ratio, resulting in conductivity thresholds at 5% carbon nanotubes [22]. On the other hand, chemical functionalization of the carbon nanotube surface with amino groups allows the percolation characteristics to be maintained; however, a lower conductivity was observed than that of the chemically untreated carbon nanotubes, obtaining a percolation threshold of 0.25% [23].

In view of the small amount of available literature regarding the effect of multiwalled carbon nanotubes modified with amine groups on the electrical and mechanical properties of thermoset polymers, as well as the many discrepancies in the currently reported results, we present research on the influence of amino-modified carbon nanotubes on the mentioned properties. Additionally, this paper presents the effect of nanoadditives obtained by the developed synthesis method on two pure epoxy resins and resins with additives used in industry, ensuring improved processing.

In this work, we present the structure, mechanical, and electrical properties of epoxy composites modified with multiwalled carbon nanotubes. Multiwalled carbon nanotubes were obtained by catalytic chemical vapor deposition, subjected to purification processes and covalent functionalization by depositing functional groups on their surfaces. The study included the analysis of the change in DC resistivity, tensile strength, strain, and Young's modulus with the addition of carbon nanotubes in the range of 0 to 2.5 wt.%. The effect of agents causing an increase in the affinity of the nanomaterial to the polymer on the abovementioned properties was also investigated.

2. Materials and Methods

2.1. Functionalization of Carbon Nanotubes

Nanocyl NC7000 multiwalled carbon nanotubes were used as the filler for the polymer matrix. Carbon nanotubes were prepared using the catalytic chemical vapor deposition (CCVD) method.

The carbon nanotubes were subjected to purification from impurities in the form of amorphous carbon, soot, and residues of catalysts used in their production process. Purification was carried out using the wet method with a mixture of oxidizing acids, i.e., concentrated nitric acid (V) and concentrated sulfuric acid (VI) in a mass ratio of 3:1. For this purpose, the multiwalled carbon nanotubes were placed in a round-bottom flask and filled with a mixture of acids. The entire mixture was subjected to ultrasonic shaking for 60 min to break up the agglomerates. After shaking the mixture using ultrasound, the nanotubes were quenched in a round-bottom flask under a reflux condenser for a period of 16 h at the boiling point of the oxidizing acid mixture.

Then, the mixture of multiwalled carbon nanotubes in oxidizing acids was filtered on filter paper under reduced pressure and rinsed with distilled water to remove residual acids until the pH value of the filtrate was 6.0. After neutralization of the acid reaction, the carbon nanotubes located on the filter paper were collected and further dispersed in isopropyl alcohol, and the mixture was filtered again. In the final stage, the rinsed carbon nanotubes were dried in a chamber dryer at 50–70 °C for 6 h.

The dried multiwalled carbon nanotubes were mixed with ammonia using a high-efficiency paddle mixer. The entire mixture was then subjected to shaking for 60 min to break up the agglomerate using ultrasound. After ultrasonic shaking, the mixture was left for approximately 24 h. Later, the mixture was filtered again through filter paper, and the collected carbon nanotubes were rinsed several times with demineralized water followed by isopropyl alcohol or ethanol. The resulting precipitate from the filter was collected and placed in an oven for approximately 6–8 h at 60 °C. The dried precipitate was subjected to crushing and grinding.

2.2. Preparation of Nanocomposites Based on Epoxy Resins with Carbon Nanotubes

The polymer nanocomposite was obtained using a low viscosity epoxy resin prepared from a mixture of 2,2-bis[4-(2,3-epoxypropoxy)phenyl]propane and 2,3-epoxypropyl o-tolyl ether (Epidian 601, CIECH Sarzyna S.A., Nowa Sarzyna, Poland). The second formulation was based on a mixture obtained from bisphenol AF, epichlorohydrin, and alkyl glycidyl ether (Epidian 652, CIECH Sarzyna S.A., Nowa Sarzyna, Poland). Both resins, hereafter designated 601 and 652, respectively, were crosslinked with isophorone diamine with salicylic acid in benzyl alcohol at a ratio of 1:2 relative to resin 601 (or 652). To obtain better dispersion, compounds improving wettability based on polyamide salts and (trimethoxysilyl)propylamine dispersed in a mixture of 2-butoxyethanol and methanol were used. The aforementioned substances were added to mixture 601 or 652 at a described amount of 1.0–3.0 wt.% relative to the total weight of the composite to improve the deaeration of the mixture. A mixture of polydimethylsiloxanes, cyclosiloxanes, and silica was used as an antifoaming agent. Composites containing additives to improve processability and the degree of dispersing the nanoadditive were designated 652* and 601*.

Using a high-speed nonaerating mixer, the previously mentioned substances were mixed, and multiwalled carbon nanotubes were added at an amount of 0.5 to 2.5 wt.% relative to the total weight of the composite while continuously mixing. Then, after the addition of the hardener, the mixture was mixed for approximately 1–2 min. Afterwards, the mixture was transferred to a vacuum mixer with a helical blade and mixed for another 5 min at maximum speed while thermostating the tank with room temperature water and a vacuum of 0.1 bar. Once all the components of the composite were thoroughly mixed, the material was poured into a mold where crosslinking was carried out. Once the gel point was reached, the molds were transferred to a chamber dryer where crosslinking of the composite was continued at elevated temperature for 8 h.

2.3. Structural Analysis of Carbon Nanotubes

Structural analyses were conducted using a JEOL scanning electron microscope (SEM) (model JSM-7800F, JEOL Ltd., Tokyo, Japan). The JEOL JSM-7800F microscope is a high-resolution scanning electron microscope equipped with four detectors: upper electron detector (UED), lower electron detector (LED), backscatter electron detector (BED), and transmission electron detector (TED).

A small amount of the analyzed carbon nanotubes was dispersed in isopropyl alcohol, then the suspension prepared using this method was spotted onto a so-called grid and left under clean conditions to dry. After complete drying, the grid was placed in a holder in the chamber of the microscope. The analysis was performed using the TED detector.

2.4. Conductivity Measurements

Material samples were prepared by casting in a cylinder mold. After release, the faces of the samples were ground using fixed sandpaper and a bench drill spindle as a drive. The samples prepared in this way were covered on both sides (parallel surfaces—cylinder bases) with conductive varnish based on silver additive (Electon 40AC, manufactured by Amepox company). To dry and harden the coating, the samples were left at a temperature of 25 degrees for 24 h. After this time, the quality of the surface conductivity was checked by applying the leads of the multimeter at the maximum distance from each other to one of the painted surfaces. The measured values did not exceed 1 Ohm. The prepared samples were placed in the developed measuring holder. It consisted of flat jaws pressed by a spring to the samples. The contacting surfaces were covered by a conductor in the form of copper sheets connected with electrical leads. As a result, a repeatable jaw pressure of approximately 24.5 N was obtained.

The prepared samples were geometrically measured using a mechanical caliper with accuracy to 2 decimal places, which gave a basis for the cross-sectional area and length calculation. A Keysight 34461A Digital Multimeter was used to measure sample resistance. The instrument's 1 h warm-up procedure and 4-wire configuration were applied to compensate for interferences and the influence of the series resistance of the measurement circuit.

2.5. Measurements of Mechanical Properties

Determination of the static tensile properties was conducted according to PN-EN ISO 527:1998 Plastics: Determination of tensile properties [24]. According to the requirements of the standard, the test sample (molding) subjected to the test is flat and has the shape of a paddle as shown in Figure 2. The dimensions of the sample are assumed to be as follows: thickness of 4.0 ± 0.2 mm, measured section width of 10 ± 0.2 mm, and total length over 150 mm. Table 1 shows all the dimensions of the sample used in the study, and Figure 3 shows photographs of the prepared molding.

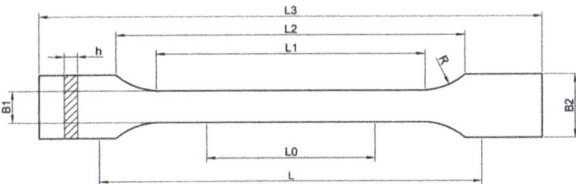

Figure 2. Dimensions of molding.

Table 1. Dimensions of molding for tensile property testing according to the EN-ISO 527-1 standard.

Dimensions of Molding	Type B1
L3—total length	150 mm
L1—length of the designated section	40 mm
R—radius	60 mm
L2—distance between parallel sections	106 mm
B2—width at ends	20 mm
B1—width at narrow section	10 mm
H—recommended thickness	4 mm
L0—measured length	50 mm
L—initial distance between handles	115 mm

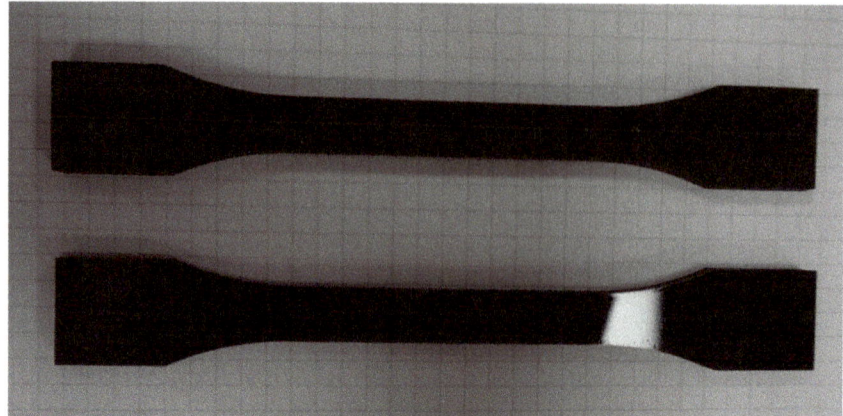

Figure 3. Photographs of moldings for tensile property testing.

The tensile property test was carried out using a high precision Shimadzu AG-X Plus 10 kN testing machine equipped with a 10 kN class 1 1/500 load cell with an initial force of 50 N and a crosshead speed of 10 mm/min (Shimadzu Corporation, Tokyo, Japan).

3. Results and Discussion

3.1. Characterizations of Functionalized CNTs

Figure 3 shows images of functionalized multiwalled carbon nanotubes.

Functionalization of multiwalled carbon nanotubes according to the described procedure allowed the formation of covalent bonds between the functional group and the sidewalls forming the nanotube cylinder [11]. The use of nanotube purification in mixed oxidizing acids resulted in many defect sites on the walls of carbon nanotubes. Defects induced by contact of a strong oxidizing compound with the nanotube are usually stabilized by bonds with carboxyl and/or hydroxyl functional groups, and their presence gives rise to further chemical reactions, including alkylation or arylation, silanization, thiolation, esterification, amidation, and grafting of polymers or biomolecules [1,5].

3.2. Measurements of Mechanical Properties

The study includes the analysis of tensile strength, strain, and Young's modulus. The lack of results for composites based on 601 resin containing 2.5 wt.% carbon nanotubes is due to the lack of composite samples. The lack of sample is caused by the significant increase in dynamic viscosity of resins in the uncrosslinked state, and the inability to pour the material into a mold. Figures 4–6 show the summary of the results obtained.

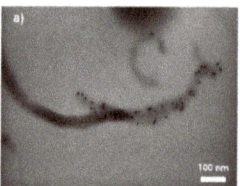 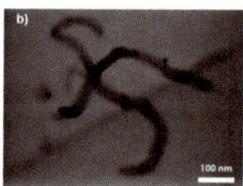 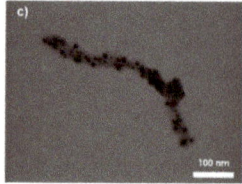

Figure 4. Transmission electron microscopy images of multiwalled carbon nanotubes with deposited amino groups: (**a**) 130,000 times magnification, (**b**) 190,000 times magnification, and (**c**) 200,000 times magnification.

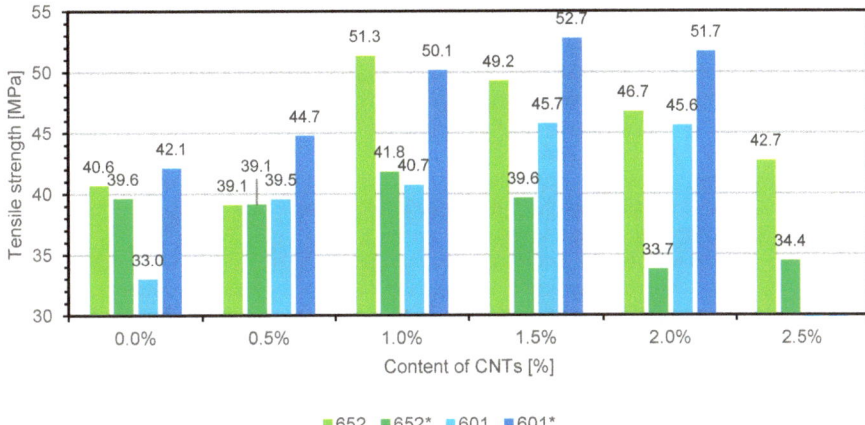

Figure 5. Chart presenting the relationship between the tensile strength of composites with different contents of multiwalled carbon nanotubes (MWCNTs).

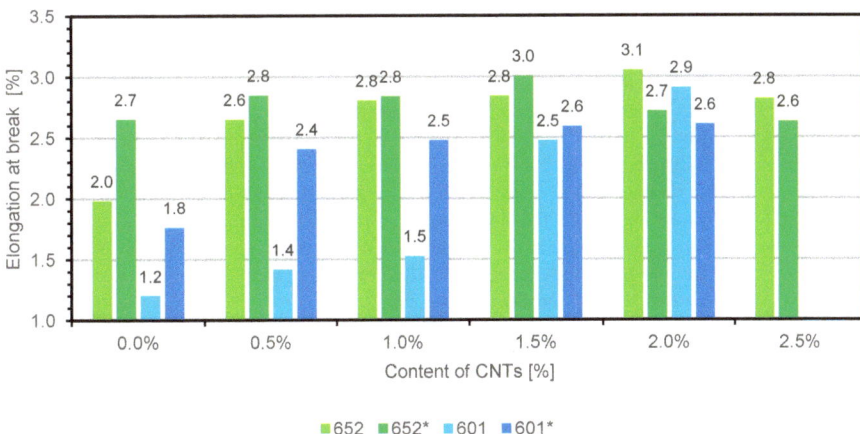

Figure 6. Chart presenting the relationship between the strain and type of composite with various contents of multiwalled carbon nanotubes (MWCNTs).

The analysis of the obtained results indicates that the addition of multiwalled carbon nanotubes in the epoxy composite improves the tensile strength. An increase in tensile strength was observed in composite 652, and the highest values were observed in the sample containing 1 wt.% of carbon nanotubes, which improved the parameter by 26% compared to the reference sample. A higher percentage of filler did not further improve the results obtained; in contrast, a gradual decrease in strength was observed, while 2.5 wt.% of nanotubes did not result in a worse result than the unfilled polymer. An improvement in the properties of polymeric materials can be achieved by filling the material, but after exceeding a certain limit, deterioration of the analyzed parameters is often observed.

Concerning composite 652*, one might note that it has a lower tensile strength than composite 652, and the highest result was obtained when the composite contained 1 wt.% multiwalled nanotubes. The higher proportion of nanofiller resulted in a deterioration of strength, obtaining worse results than the reference sample, which may indicate a high degree of agglomeration of carbon nanotubes.

The use of carbon nanotubes in composite 601 also resulted in the strengthening of the polymeric matrix, and this trend was observed until reaching the maximum strength at 1.5 wt.% filling of the polymeric matrix. Comparable results were obtained for composite 601*, where the addition of substances improving the wettability of the nanoadditive surface ensured better strength parameters.

Figure 5 shows the relationship between the strain and the type of composites with various contents of multiwalled carbon nanotubes (MWCNTs).

The analysis of the above chart shows that, with increasing filler content, the percentage strain of the material increases, reaching an extreme value when the characteristic amount of the additive is used. For composites 652 and 652*, the extremes are reached at 2.0 wt.% and 1.5 wt.% carbon nanotube proportions, respectively, while for composites 601 and 601*, the highest strain values are obtained at the 2 wt.% matrix fill. The use of wetting agents and antifoaming agents based on silicone compounds influenced the increase in strain, which was caused by the higher mobility of polymer chains, which is particularly evident in the results obtained for composites that do not contain nanomaterials. Figure 6 shows the relationship between Young's modulus and the type of composites with various contents of multiwalled carbon nanotubes (MWCNTs).

The application of multiwalled carbon nanotubes as polymer fillers in epoxy composites alters the Young's modulus. Deterioration was observed in composites 652 and 652* after the addition of 0.5 wt.% multiwalled nanotubes, although properties similar to, or slightly better than, those of the reference sample were obtained at higher proportions of the nanoadditive. The highest Young's modulus values were obtained for composite 652 containing 1.5 wt.% carbon nanotubes, while for composite 652*, 1 wt.% carbon nanotubes allowed higher Young's modulus values to be obtained. Higher contents led to a deterioration of this property. Composites containing modifying additives generally caused the value of Young's modulus to decrease.

Concerning epoxy 601, when the material was filled with carbon nanotubes up to 1.0 wt.%, no significant change in Young's modulus was observed. A higher content of nanostructures in the composite caused Young's modulus values to decrease. A comparable result was observed for epoxy composite 601*, noting that the decrease was observed only when the composite contained 2% carbon nanotubes.

In the case of composite 601, a 20% increase in tensile strength and a 1.3% increase in Young's modulus value were observed with the addition of 0.5 wt.% carbon nanotubes. To a certain concentration limit, a higher proportion of carbon nanotubes resulted in an increase in tensile strength with a gradual deterioration of Young's modulus. Comparable correlations were observed for composite 601*.

When measuring composite 652, it was established that, due to the use of 1.5% carbon nanotubes, there was a 26.3% increase in tensile strength and a 1.6% increase in Young's modulus value. For composite 652*, a 1.0% addition enabled a 4% improvement in mechanical properties, with a 4% increase in Young's modulus value and a 5.6% increase in tensile strength. Table 2 summarizes the results obtained in the studies described in relation to those reported in previous literature reports.

Table 2. Mechanical properties of pure epoxy and CNT/epoxy composites.

CNT Type	CNT WeightFraction	Young's Modulus		Tensile Strength		Ref.
	[%]	[GPa]	ΔE [%]	[MPa]	Δσ [%]	
652 with no filler	0.0	3.28	-	40.64	-	
MWCNT-NH$_2$	0.5	2.64	−19.5	39.06	−3.9	
	1.0	3.33	1.6	51.31	26.3	
	1.5	3.37	2.6	49.24	21.2	
	2.0	2.99	−8.9	46.73	15.0	
	2.5	2.89	−11.9	42.67	5.0	
652* with no filler	0.0	2.84	-	39.58	-	
MWCNT-NH$_2$	0.5	2.70	−4.7	39.12	−1.2	
	1.0	2.95	4.0	41.79	5.6	
	1.5	2.65	−6.5	39.61	0.1	
	2.0	2.55	−10.3	33.74	−14.8	Present work
	2.5	2.64	−7.0	34.44	−13.0	
601 with no filler	0.0	3.80	-	32.96	-	
MWCNT-NH$_2$	0.5	3.85	1.3	39.54	20.0	
	1.0	3.80	−0.1	40.71	23.5	
	1.5	3.27	−13.8	45.73	38.8	
	2.0	3.40	−20.8	45.59	38.8	
	2.5	-	-	-	38.3	
601* with no filler	0.0	3.64	-	42.07	-	
MWCNT-NH$_2$	0.5	3.51	−3.6	44.74	6.3	
	1.0	3.50	−3.8	50.13	19.1	
	1.5	3.48	−4.4	52.73	25.3	
	2.0	-	-	51.67	22.8	
	2.5	-	-	-	-	
Epoxy with no filler	0.0	1.48	-	46.46		
MWCNT	0.5	1.68	13.5	50.25	8.2	[25]
	1.0	1.87	26.4	58.65	26.2	
	3.0	1.69	14.2	54.48	17.3	
Epoxy with no filler	0.0	2.60	-	63.80		
MWCNT	0.1	2.78	6.9	62.97	−1.3	[14]
	0.3	2.77	6.5	63.17	−1.0	
	0.5	2.61	0.4	61.52	−3.6	
Epoxy with no filler	0.0	2.60		63.80		
MWCNT-NH$_2$	0.1	2.88	10.8	64.67	1.4	[14]
	0.3	2.81	8.1	63.64	−0.3	
	0.5	2.82	8.5	64.27	0.7	
Epoxy with no filler	0.0	1.21	-	26.00	-	
Untreated CNTs	1.0	1.38	14.0	42.00	61.5	
Acid treated CNTs	1.0	1.22	0.8	44.00	69.2	[26]
Amine treated CNTs	1.0	1.23	1.7	47.00	80.8	
Plasma treated CNTs	1.0	1.61	33.1	58.00	123.1	
Epoxy with no filler	0.0	1.18	-	52.00	-	
MWCNT	2.0	1.18	0.0	46.00	−11.5	
MWCNT-PVA	2.0	1.35	14.4	55.00	5.8	[27]
MWCNT-AEO9	2.0	1.39	17.8	62.00	19.2	
MWCNT-AEO7	2.0	1.34	13.6	58.00	11.5	

The obtained results are consistent with the currently available scientific literature data on the use of carbon nanotubes as reinforcement in epoxy composites [14,25–28]. The results confirm that the addition of multiwalled carbon nanotubes can improve mechanical properties, including tensile strength and Young's modulus, while increasing the plasticity of the composite under loading [5]. Such dependence is observed up to a certain limit value, and then a gradual or rapid deterioration of the mentioned parameters is observed [29]. In the present work, a significant improvement in tensile strength and a several percent improvement in Young's modulus were obtained. The obtained results are slightly better than those reported by Gojny et al. [14] with the addition of 0.5 wt.%. MWCNTs (tests were carried out only with carbon nanotube contents of 0.1%, 0.3%, and 0.5%). In most of the literature, the addition of carbon nanotubes has a greater effect on the change in Young's modulus than the change in tensile strength [11], and parameters obtained through the use of aminofunctionalized multiwalled carbon nanotubes did not give better results than those of previously reported pristine multiwall modified carbon nanotubes used as a reinforcing additive [30].

The increase in tensile strength of polymer composites with 1D nanoadditives results from the transfer of loads occurring in the material to the reinforcing phase. Considering carbon nanotubes, this effect is caused by a large surface area and covalent bond formation between polymer chains. The reinforcement effect can be achieved by arranging polymer chains along the carbon nanotubes in the axial direction. This method of arranging the filler may be achieved by the presence of amine groups on the carbon nanotube surface, which leads to a reduction in the average length of the polymer chain, as well as an improvement in the cross-linking of polymer chains. As a result, an increase in tensile strength and longitudinal tensile modulus can be achieved.

The analysis of the obtained results shows that no significant changes in the Young's modulus are observed. In the case of the tested composites, the condensation polymerization process was carried out until a specific viscosity of resin was achieved. At the same time, the shortening of the carbon nanotube structure caused by chemical functionalization reduced the interface between the polymer matrix and the dispersed phase, which resulted in no improvement or even deterioration of the elastic modulus.

3.3. Results of Direct Current Resistivity Measurements

The resistivity is measured using an indirect method by measuring the vertical resistivity, taking into account the surface area and thickness of the sample. In this study, the resistivity of cylindrical samples with dimensions of approximately 10 mm in height and approximately 12 mm in base diameter was measured. The measurements of the resistivity of the samples were recalculated based on their exact dimensions, thus obtaining the resistivity values, which are summarized in the chart below depending on the mass fraction of carbon nanotubes in the finished composite. Two measurements were performed for each sample, and the mean value is presented (Figure 7).

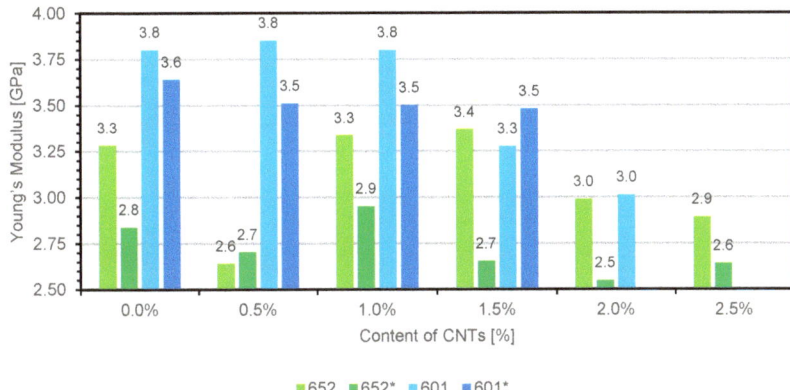

Figure 7. Chart presenting the relationship between Young's modulus and the type of composites with various contents of multiwalled carbon nanotubes (MWCNTs).

The study indicates that the addition of multiwalled carbon nanotubes reduces the resistivity of the material, thus improving its conductivity (Figure 8). For composites not containing carbon nanotubes, no results were obtained as, due to their dielectric properties, they were beyond the measurement range of the device. For all the epoxy composites tested, resistivity results were recorded at a 1 wt.% content of multiwalled carbon nanotubes, with 1.5 percent nanoadditive causing a significant increase in electrical conductivity. The addition of large amounts of carbon nanotubes to the polymer matrix results in further improvement in conductivity, reaching a plateau.

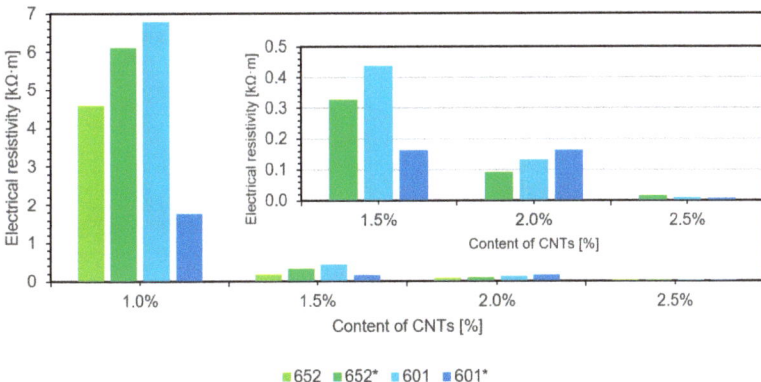

Figure 8. Chart presenting the relationship between the resistivity and weight fraction of multiwalled carbon nanotubes for the composites studied.

Comparative studies indicate that additives improving the processing and dispersion of carbon nanotubes worsened the electrical conductivity in composites based on sample 652. The opposite effect was obtained for composite 601, therefore one might assume that the selected components for this polymer matrix improved the degree of dispersion of the carbon additive. The greatest effect was obtained by the additive in composite 601. The measured resistivity is generally lower in composites based on 601, which may be related to the higher ability to form crosslinks that allow charge flow. Table 3 summarizes the results obtained in the studies, described in relation to those reported in previous literature reports.

Table 3. Electrical conductivity of the nanocomposites as a function of filler content in weight percent.

CNT Type	Content of CNTs [%]	Processing Method	Electrical Resistivity [Ω·m]	Electrical Conductivity [S/m]	Ref.
652 with no filler	0.0		-	-	
MWCNT-NH$_2$	0.5	Vacuum stirred	-	-	
	1.0		4.60×10^3	2.18×10^{-4}	
	1.5		1.70×10^2	5.88×10^{-3}	
	2.0		7.35×10^1	1.36×10^{-2}	
	2.5		9.43×10^0	1.06×10^{-1}	
652* with no filler	0.0		-	-	
MWCNT-NH$_2$	0.5	Vacuum stirred	-	-	
	1.0		6.11×10^3	1.64×10^{-4}	
	1.5		3.28×10^2	3.05×10^{-3}	
	2.0		9.23×10^1	1.08×10^{-2}	
	2.5		1.62×10^1	6.18×10^{-2}	Present work
601 with no filler	0.0		-	-	
MWCNT-NH$_2$	0.5	Vacuum stirred	-	-	
	1.0		6.78×10^3	1.47×10^{-4}	
	1.5		4.37×10^2	2.29×10^{-3}	
	2.0		1.33×10^2	7.52×10^{-3}	
	2.5		-	-	
601* with no filler	0.0		-	-	
MWCNT-NH$_2$	0.5	Vacuum stirred	-	-	
	1.0		1.77×10^3	5.66×10^{-4}	
	1.5		1.63×10^2	6.13×10^{-3}	
	2.0		1.63×10^2	6.13×10^{-3}	
	2.5		8.62×10^0	1.16×10^{-1}	
MWCNT	1.0	Stirred, heat sheared (slowly)	2.50×10^0	4.00×10^{-1}	
MWCNT	1.0	Stirred, heat sheared (medium)	3.33×10^0	3.00×10^{-1}	[21]
MWCNT	0.6	Stirred, heat sheared (fast)	2.50×10^1	4.00×10^{-2}	
MWCNT	0.3	Calendered, stirred	1.00×10^2	1.00×10^{-2}	
MWCNT NH$_2$-functionalized	0.4	Calendered, stirred	2.00×10^3	5.00×10^{-4}	[31]
MWCNTs	0.5	Calendered, stirred	1.00×10^2	1.00×10^{-2}	
MWCNT	1.0	Sonicated	5.00×10^0	2.00×10^{-1}	[28]
MWCNT	1.0	Stirred	5.00×10^1	2.00×10^{-2}	
MWCNTs	10.0	Solution mixing	3.33×10^2	3.00×10^{-3}	[32]
SDS suspended MWCNTs	0.5	Bulk mixing	4.00×10^6	2.50×10^{-7}	[33]
MWCNTs	2.5	Solution mixing	7.69×10^1	1.30×10^{-2}	[34]
MWCNTs	1.4	Solution mixing	2.00×10^0	5.00×10^{-1}	[35]
Oxidized MWCNTs	1.0	Solution mixing	1.00×10^2	1.00×10^{-2}	[36]
Pristine MWCNTs	1.0	Solution mixing	1.00×10^2	1.00×10^{-2}	[37]
MWCNTs	5.0	Calendered, stirred	2.00×10^{-2}	5.00×10^1	[38]
MWCNT	1.0	Heat sheared	5.00×10^{-1}	2.00×10^0	[20]
MWCNT	0.5	Heat sheared	2.50×10^0	4.00×10^{-1}	[39]
MWCNT	3.0	Stirred	2.00×10^{-1}	5.00×10^0	[40]
MWCNT	10.0	Stirred	2.00×10^2	5.00×10^{-3}	[32]
MWCNT	4.0	Stirred, hot pressed	2.00×10^{-1}	5.00×10^0	[41]
MWCNT	2.0	Calendered, vacuum stirred	3.33×10^1	3.00×10^{-2}	[42]

The obtained results are consistent with the currently available literature data on the use of carbon nanotubes in epoxy composites [20,21,28,31–37], while for nanotubes functionalized with amines, slightly worse results than those reported were obtained [23]. The surface functionalization of carbon nanotubes is crucial concerning the electrical conductivity of polymer nanocomposites by facilitating the dispersion of the additive and the formation of uniformly distributed conductive crosslinks, which ultimately lowers the percolation threshold. However, it is worth noting that excessive modification introducing many heterogeneous atoms on the surface causes electron flow disturbances, thus degrading the electrical properties resulting from the addition of carbon nanotubes alone [1]. The lower values of measured electrical conductivities are probably caused by chemical treatment with strong oxidizing acids and the use of ultrasound, which shortens the length of carbon nanostructures through their excessive degradation. Purification and functionalization based on the use of strongly oxidizing acids, e.g., sulfuric acid or nitric acid, can cause degradation of the structure and consequently, their shortening and change in the length-to-diameter ratio. Both of these changes adversely affect the electrical conductivity obtained [42].

The conclusions drawn are confirmed by the illustrative photos in Figure 4, where the lengths of the nanotubes are smaller than the dimensions of the structures declared by the manufacturer before the functionalization process. The reduction of the aspect ratio made it necessary to use a higher mass fraction of carbon filler to obtain a conductive network in the polymer matrix with dielectric properties. To obtain the same order of electrical conductivity values as those obtained with 1 wt.% addition of carbon nanotubes in the works of [21,28,32], it was necessary to use 2.5 wt.% carbon nanotubes functionalized with amine groups.

4. Conclusions

This paper presents the effect of multiwalled carbon nanotubes containing amine groups on their surfaces on the electrical and mechanical properties of epoxy composites. The addition of carbon nanotubes allowed us to obtain an electrically conductive material, in which nanotubes form a network for the charge flowing through the material. For all materials, the percolation threshold was obtained with 1% addition of multiwalled carbon nanotubes, and filling the polymer with a higher content of carbon nanotubes increased its conductivity. Carbon nanotube purification in a mixture of oxidizing acids strongly defected the nanotube structure. This allowed the formation of multiple bonds at the defect sites, allowing the grafting of functional groups, thereby increasing the compatibility with the polymer matrix and obtaining a better degree of dispersion. However, the defect may have contributed to shortening the length of the carbon nanotubes, which resulted in poorer percolation thresholds with respect to the literature data.

The use of carbon nanotubes as polymer reinforcement allows higher values of tensile strength and a higher strain percentage to be achieved. In contrast, Young's modulus values did not increase significantly, and higher nanofiller percentages resulted in a drastic decrease in the values of the abovementioned properties.

An increase in tensile strength was achieved when the weight fraction of nanotubes did not exceed 1.5 wt.% for composite 652* and 2.0 wt.% for composites 652, 601, and 601*. The addition of larger amounts of carbon filler caused the deterioration of properties, and the achieved effect is characteristic for polymer composites where nanomaterials of linear and layered structures were added. This is due to the formation of aggregates and agglomerates of carbon nanotubes, which may result in a significant increase in viscosity of the uncross-linked resin, and thus hinder dispersion of the additive in the polymer (macroscopic dispersion) and disentanglement of carbon nanotubes (nanoscopic dispersion). The compounds proposed can be used to obtain a material with uniformly dispersed carbon nanotubes.

Due to the difficulty in achieving a high degree of homogenization of carbon nanotubes in an epoxy matrix, the strength parameters of the composites are often lower than expected.

Author Contributions: Conceptualization, P.S. and T.C.; methodology, P.S.; validation, P.S., T.C., and A.R.; formal analysis, P.S.; investigation, P.S., Z.K., D.G. and M.O.-P.; resources, P.S.; data curation, Z.K., D.G. and M.O.-P.; writing—original draft preparation, P.S.; writing—review and editing, T.C. and A.R.; visualization, P.S.; supervision, T.C. and A.R.; project administration, T.C.; and funding acquisition, T.C. All authors have read and agreed to the published version of the manuscript.

Funding: This research was funded by the statutory sources of the Department of Materials Technology, Military University of Technology.

Institutional Review Board Statement: Not applicable.

Informed Consent Statement: Not applicable.

Data Availability Statement: The data underlying this article will be shared on reasonable request from the corresponding author.

Conflicts of Interest: The authors declare no conflict of interest. The funders had no role in the design of the study; in the collection, analyses, or interpretation of data; in the writing of the manuscript, or in the decision to publish the results.

References

1. Ma, P.C.; Siddiqui, N.A.; Marom, G.; Kim, J.K. Dispersion and functionalization of carbon nanotubes for polymer-based nanocomposites: A review. *Compos. Part A Appl. Sci. Manuf.* **2010**, *41*, 1345–1367. [CrossRef]
2. Bhattacharya, M. Polymer Nanocomposites—A Comparison between Carbon Nanotubes, Graphene, and Clay as Nanofillers. *Materials* **2016**, *9*, 262. [CrossRef] [PubMed]
3. Królikowski, W.; Rosłaniec, Z. Nanokompozyty polimerowe. *Kompozyty* **2004**, *4*, 3–15.
4. Mittal, G.; Dhand, V.; Rhee, K.Y.; Park, S.J.; Lee, W.R. A review on carbon nanotubes and graphene as fillers in reinforced polymer nanocomposites. *J. Ind. Eng. Chem.* **2015**, *21*, 11–25. [CrossRef]
5. Spitalsky, Z.; Tasis, D.; Papagelis, K.; Galiotis, C. Carbon nanotube-polymer composites: Chemistry, processing, mechanical and electrical properties. *Prog. Polym. Sci.* **2010**, *35*, 357–401. [CrossRef]
6. Huczko, A.; Kurcz, M.; Popławska, M. *Nanorurki Węglowe. Otrzymywanie, Charakterystyka, Zastosowania*; Wydawnictw: Warszawa, Poland, 2014.
7. Kausar, A.; Rafique, I.; Muhammad, B. Significance of Carbon Nanotube in Flame-Retardant Polymer/CNT Composite: A Review. *Polym. Plast. Technol. Eng.* **2017**, *56*, 470–487. [CrossRef]
8. Smart, S.K.; Cassady, A.I.; Lu, G.Q.; Martin, D.J. The biocompatibility of carbon nanotubes. *Carbon* **2006**, *44*, 1034–1047. [CrossRef]
9. Xie, X.L.; Mai, Y.W.; Zhou, X.P. Dispersion and alignment of carbon nanotubes in polymer matrix: A review. *Mater. Sci. Eng. R Rep.* **2005**, *49*, 89–112. [CrossRef]
10. Kim, S.W.; Kim, T.; Kim, Y.S.; Choi, H.S.; Lim, H.J.; Yang, S.J.; Park, C.R. Surface modifications for the effective dispersion of carbon nanotubes in solvents and polymers. *Carbon* **2012**, *50*, 3–33. [CrossRef]
11. Domun, N.; Hadavinia, H.; Zhang, T.; Sainsbury, T.; Liaghat, G.H.; Vahid, S. Improving the fracture toughness and the strength of epoxy using nanomaterials-a review of the current status. *Nanoscale* **2015**, *7*, 10294–10329. [CrossRef]
12. Andrews, R.; Weisenberger, M.C. Carbon nanotube polymer composites. *Curr. Opin. Solid State Mater. Sci.* **2004**, *8*, 31–37. [CrossRef]
13. Coleman, J.N.; Khan, U.; Blau, W.J.; Gun'ko, Y.K. Small but strong: A review of the mechanical properties of carbon nanotube-polymer composites. *Carbon* **2006**, *44*, 1624–1652. [CrossRef]
14. Gojny, F.H.; Wichmann, M.H.G.; Fiedler, B.; Schulte, K. Influence of different carbon nanotubes on the mechanical properties of epoxy matrix composites—A comparative study. *Compos. Sci. Technol.* **2005**, *65*, 2300–2313. [CrossRef]
15. Špitalský, Z.; Matějka, L.; Šlouf, M.; Konyushenko, E.N.; Kovářová, J.; Zemek, J.; Kotek, J. Modification of carbon nanotubes and its effect on properties of carbon nanotube/epoxy nanocomposites. *Polym. Compos.* **2009**, *30*, 1378–1387. [CrossRef]
16. Guo, P.; Chen, X.; Gao, X.; Song, H.; Shen, H. Fabrication and mechanical properties of well-dispersed multiwalled carbon nanotubes/epoxy composites. *Compos. Sci. Technol.* **2007**, *67*, 3331–3337. [CrossRef]
17. Breton, Y.; Désarmot, G.; Salvetat, J.P.; Delpeux, S.; Sinturel, C.; Béguin, F.; Bonnamy, S. Mechanical properties of multiwall carbon nanotubes/epoxy composites: Influence of network morphology. *Carbon* **2004**, *42*, 1027–1030. [CrossRef]
18. Bauhofer, W.; Kovacs, J.Z. A review and analysis of electrical percolation in carbon nanotube polymer composites. *Compos. Sci. Technol.* **2009**, *69*, 1486–1498. [CrossRef]
19. Kaur, G.; Adhikari, R.; Cass, P.; Bown, M.; Gunatillake, P. Electrically conductive polymers and composites for biomedical applications. *RSC Adv.* **2015**, *5*, 37553–37567. [CrossRef]
20. Sandler, J.K.W.; Kirk, J.E.; Kinloch, I.A.; Shaffer, M.S.P.; Windle, A.H. Ultra-low electrical percolation threshold in carbon-nanotube-epoxy composites. *Polymer* **2003**, *44*, 5893–5899. [CrossRef]
21. Kovacs, J.Z.; Velagala, B.S.; Schulte, K.; Bauhofer, W. Two percolation thresholds in carbon nanotube epoxy composites. *Compos. Sci. Technol.* **2007**, *67*, 922–928. [CrossRef]

22. Pécastaings, G.; Delhaès, P.; Derré, A.; Saadaoui, H.; Carmona, F.; Cui, S. Role of Interfacial Effects in Carbon Nanotube/Epoxy Nanocomposite Behavior. *J. Nanosci. Nanotechnol.* **2004**, *4*, 838–843. [CrossRef] [PubMed]
23. Kovacs, J.Z.; Andresen, K.; Pauls, J.R.; Garcia, C.P.; Schossig, M.; Schulte, K.; Bauhofer, W. Analyzing the quality of carbon nanotube dispersions in polymers using scanning electron microscopy. *Carbon* **2007**, *45*, 1279–1288. [CrossRef]
24. International Organization for Standardization. Plastics—Determination of tensile properties—Part 1: General principles. In *International Organization for Standardization. ISO 527-1:1993*; ISO: Geneva, Switzerland, 1993. Available online: https://www.iso.org/standard/4592.html (accessed on 15 June 2021).
25. Zakaria, M.R.; Abdul Kudus, M.H.; Akil, H.M.; Thirmizir, M.Z. Comparative study of graphene nanoparticle and multiwall carbon nanotube filled epoxy nanocomposites based on mechanical, thermal and dielectric properties. *Compos. Part B Eng.* **2017**, *119*, 57–66. [CrossRef]
26. Kim, J.A.; Seong, D.G.; Kang, T.J.; Youn, J.R. Effects of surface modification on rheological and mechanical properties of CNT/epoxy composites. *Carbon* **2006**, *44*, 1898–1905. [CrossRef]
27. Ying, Z.; Du, J.-H.; Bai, S.; Li, F.; Liu, C.; Cheng, H.-M. Mechanical properties of surfactant-coating carbon nanofiber/epoxy composite. *Int. J. Nanosci.* **2002**, *01*, 425–430. [CrossRef]
28. Li, J.; Ma, P.C.; Chow, W.S.; To, C.K.; Tang, B.Z.; Kim, J.-K. Correlations between Percolation Threshold, Dispersion State, and Aspect Ratio of Carbon Nanotubes. *Adv. Funct. Mater.* **2007**, *17*, 3207–3215. [CrossRef]
29. Montazeri, A.; Montazeri, N. Viscoelastic and mechanical properties of multi walled carbon nanotube/epoxy composites with different nanotube content. *Mater. Des.* **2011**, *32*, 2301–2307. [CrossRef]
30. Park, J.M.; Kim, D.S.; Lee, J.R.; Kim, T.W. Nondestructive damage sensitivity and reinforcing effect of carbon nanotube/epoxy composites using electro-micromechanical technique. *Mater. Sci. Eng. C* **2003**, *23*, 971–975. [CrossRef]
31. Gojny, F.H.; Wichmann, M.H.G.; Fiedler, B.; Kinloch, I.A.; Bauhofer, W.; Windle, A.H.; Schulte, K. Evaluation and identification of electrical and thermal conduction mechanisms in carbon nanotube/epoxy composites. *Polymer* **2006**, *47*, 2036–2045. [CrossRef]
32. Yuen, S.-M.; Ma, C.-C.M.; Wu, H.-H.; Kuan, H.-C.; Chen, W.-J.; Liao, S.-H.; Hsu, C.-W.; Wu, H.-L. Preparation and thermal, electrical, and morphological properties of multiwalled carbon nanotube and epoxy composites. *J. Appl. Polym. Sci.* **2007**, *103*, 1272–1278. [CrossRef]
33. dos Santos, A.S.; Leite, T.d.O.N.; Furtado, C.A.; Welter, C.; Pardini, L.C.; Silva, G.G. Morphology, thermal expansion, and electrical conductivity of multiwalled carbon nanotube/epoxy composites. *J. Appl. Polym. Sci.* **2008**, *108*, 979–986. [CrossRef]
34. Barrau, S.; Demont, P.; Peigney, A.; Laurent, C.; Lacabanne, C. Dc and ac conductivity of carbon nanotubes-polyepoxy composites. *Macromolecules* **2003**, *36*, 5187–5194. [CrossRef]
35. Song, Y.S.; Youn, J.R. Influence of dispersion states of carbon nanotubes on physical properties of epoxy nanocomposites. *Carbon* **2005**, *43*, 1378–1385. [CrossRef]
36. Špitalský, Z.; Krontiras, C.A.; Georga, S.N.; Galiotis, C. Effect of oxidation treatment of multiwalled carbon nanotubes on the mechanical and electrical properties of their epoxy composites. *Composites Part A Appl. Sci. Manuf.* **2009**, *40*, 778–783. [CrossRef]
37. Liu, L.; Etika, K.C.; Liao, K.-S.; Hess, L.A.; Bergbreiter, D.E.; Grunlan, J.C. Comparison of Covalently and Noncovalently Functionalized Carbon Nanotubes in Epoxy. *Macromol. Rapid Commun.* **2009**, *30*, 627–632. [CrossRef] [PubMed]
38. Thostenson, E.T.; Chou, T.W. Processing-structure-multi-functional property relationship in carbon nanotube/epoxy composites. *Carbon* **2006**, *44*, 3022–3029. [CrossRef]
39. Moisala, A.; Li, Q.; Kinloch, I.A.; Windle, A.H. Thermal and electrical conductivity of single- and multi-walled carbon nanotube-epoxy composites. *Compos. Sci. Technol.* **2006**, *66*, 1285–1288. [CrossRef]
40. Bai, J.B.; Allaoui, A. Effect of the length and the aggregate size of MWNTs on the improvement efficiency of the mechanical and electrical properties of nanocomposites - Experimental investigation. *Compos. Part A Appl. Sci. Manuf.* **2003**, *34*, 689–694. [CrossRef]
41. Wichmann, M.H.G.; Sumfleth, J.; Fiedler, B.; Gojny, F.H.; Schulte, K. Multiwall carbon nanotube/epoxy composites produced by a masterbatch process. *Mech. Compos. Mater.* **2006**, *42*, 395–406. [CrossRef]
42. Nanni, F.; Valentini, M. Electromagnetic properties of polymer-carbon nanotube composites. In *Polymer-Carbon Nanotube Composites: Preparation, Properties and Applications*; Elsevier Ltd.: Amsterdam, The Netherlands, 2011; pp. 329–346, ISBN 9781845697617.

Article

Effect of Strain on Heating Characteristics of Silicone/CNT Composites

Minoj Gnanaseelan [1,*], Kristin Trommer [1], Maik Gude [2], Rafal Stanik [2], Bartlomiej Przybyszewski [3], Rafal Kozera [3] and Anna Boczkowska [3]

1. FILK Freiberg Institute gGmbH, Meißner Ring 1-5, 09599 Freiberg, Germany; kristin.trommer@filkfreiberg.de
2. Institute of Lightweight Engineering and Polymer Technology (ILK), Technische Universität Dresden, 01307 Dresden, Germany; maik.gude@tu-dresden.de (M.G.); rafal.stanik@tu-dresden.de (R.S.)
3. Faculty of Materials Science and Engineering, Warsaw University of Technology, 141 Woloska Str., 02-507 Warsaw, Poland; Bartlomiej.przybyszewski.dokt@pw.edu.pl (B.P.); rafal.kozera@pw.edu.pl (R.K.); anna.boczkowska@pw.edu.pl (A.B.)
* Correspondence: minoj.gnanaseelan@filkfreiberg.de; Tel.: +49-3731-366-169

Citation: Gnanaseelan, M.; Trommer, K.; Gude, M.; Stanik, R.; Przybyszewski, B.; Kozera, R.; Boczkowska, A. Effect of Strain on Heating Characteristics of Silicone/CNT Composites. *Materials* **2021**, *14*, 4528. https://doi.org/10.3390/ma14164528

Academic Editor: Lei Zhai

Received: 30 June 2021
Accepted: 3 August 2021
Published: 12 August 2021

Publisher's Note: MDPI stays neutral with regard to jurisdictional claims in published maps and institutional affiliations.

Copyright: © 2021 by the authors. Licensee MDPI, Basel, Switzerland. This article is an open access article distributed under the terms and conditions of the Creative Commons Attribution (CC BY) license (https://creativecommons.org/licenses/by/4.0/).

Abstract: In this work, silicone/carbon nanotube (CNT) composites were produced using a spread coating process, followed by morphological investigations and determination of their electrical properties and heating behaviour through the application of electric potential. Composites containing varying amounts of CNT (1–7%) were investigated for their thermal behaviour with the use of an IR camera. Subsequently, thermal behaviour and electrical properties were measured when the samples were stretched (up to 20%). With the 7% CNT composites, which had a conductivity of 106 S/m, it was possible to achieve a temperature of 155 °C at a relatively low voltage of 23 V. For high CNT contents, when the potential was controlled in such a way as to maintain the temperature well below 100 °C, the temperature remained almost constant at all levels of strain investigated. At higher potentials yielding temperatures around 100 °C and above, stretching had a drastic effect on temperature. These results are critical for designing composites for dynamic applications requiring a material whose properties remain stable under strain.

Keywords: CNT composites; silicone: Joule heating; conductive polymer composites; spread coating; electrical heating

1. Introduction

As many fields of engineering focus on the development of materials that are functional, lightweight, and flexible, conductive polymer composites (CPCs) have attracted interest, as they possess these essential attributes [1]. CPCs are prepared by incorporating electrically conductive particles, such as carbon nanotubes (CNTs), graphene, carbon black, graphite, etc., into an insulating polymer matrix. The polymer matrix can be hard and inflexible, such as epoxy and polymethyl methacrylate (PMMA); tough, such as polyurethane (PU) and polycarbonate; or stretchable, such as silicone, elastomers, and thermoplastic polyurethane (TPU). CPCs have several applications, such as electromagnetic interference (EMI) shielding, antistatic, thermoelectric materials, sensors, actuators, etc. [2,3].

In this paper, CNT-reinforced silicone rubber (SR) composites were investigated. Silicone rubbers are some of the most important materials among inorganic synthetic functional elastomers. They possess unique properties and advantages, such as chemical resistance, thermal stability, low toxicity, and above all high elasticity [4,5]. Because of these properties they are widely used in different sectors, such as for medical devices, implants, sealants, electronics, lubricants, and membranes [6,7]. In practice, SR cannot be used on its own because of its low mechanical properties (tensile strength and Young's modulus) as well as its low electrical conductivity and thermal stability [8]. Consequently, different fillers are incorporated into an SR matrix to improve its mechanical as well as its functional properties [9,10].

In recent times, carbon allotropes have attracted scientists owing to their ability to improve the thermal stability and electrical conductivity of polymer matrices [11,12]. Among all electrically conductive carbon allotropes, the most preferred types are multiwalled CNTs (MWCNTs) because of their low price, accessibility, and extraordinary properties (such as their high tensile strength) [13], as well as their excellent thermal and electrical conductivity [14,15]. Several polymer composites filled with CNTs, when investigated in recent research, exhibited improved thermal and electrical conductivity. The mechanism of the improvement in properties due to CNTs has been studied [16–18]. Unlike other fillers, such as carbon black and graphite, even with a relatively low loading of CNTs into the SR matrix, the mechanical, thermal, or electrical properties of the composites may be significantly improved, making CNTs a promising candidate for a filler. The vastly improved thermal stability of CNT-reinforced plastics has been ascertained in the literature [19,20]. In studies utilising a large amount of CNT as filler, mechanical properties were improved with little deterioration in the inherent properties (e.g., elasticity) of the polymer matrix. For example, CNT/rubber composites show higher thermal stability than that of neat polymer matrix and retain good mechanical properties with only a slight reduction in elongation at break [21]. It has also been proven that the final properties of SR–CNT composites depend not only on their CNT concentration but also on their size (outer diameter, inner diameter, and length) [22]. Apart from mechanical and electrical properties, the electrical heating characteristics of SR–CNT composites have also been investigated previously, but only under static conditions [23–25]. As many applications demand stable performance even when the material is stretched, the electrical heating behaviour needs to be analysed also when strained, which has not been reported in earlier works. Hence, the main objective of this work revolves around a comparison of the heating characteristics of several composites under strain.

In this study, CNT/silicone rubber nanocomposites with improved heating behaviour and electrical conductivity were fabricated. In addition to the measurement of static heating characteristics, the effect of strain on heating behaviour as well as on electrical conductivity was investigated.

2. Materials and Methods
2.1. Chemicals

Liquid silicone rubber (LSR), Elastosil LR 6250 F from Wacker Chemie AG (Munich, Germany) with a viscosity of 100 Pa·s, shore A hardness of 36, tensile strength of 5 MPa and elongation at break of 350%, was used along with a hydrogen-terminated polysiloxane (Crosslinker W). Multiwalled CNT (MWCNT), NC7000 from Nanocyl^{TM} from Belgium (Sambreville), with an average diameter of 9.5 nm, a length of 1.5 µm and carbon purity of 90%, served as conductive filler. Toluene (99.5% purity) from Roth, and silver powder (99.5% purity) with a mean particle size of 5µm were also used. All the chemicals were used as purchased without any further purification.

2.2. Preparation of Silicone/CNT Dispersion

A defined quantity of MWCNT was added to LSR in a container, and the mixture was stirred with a spatula to ensure that the CNT powders were macroscopically mixed. Then, the mixture was introduced into a 3-roll mill, EXAKT 120EH-250, as the rolls rotated. The mixture was sheared through the rolls in 4 discrete passes with decreasing gaps between the rolls. The roll gaps are presented in Table 1.

After the final pass, the dispersion was collected from the last roll in a container. This final step played a dominant role in dispersing CNT in the silicone matrix.

Table 1. Gap between the rolls set in the 3-roll mixing process.

Sequence	1st Gap (µm)	2nd Gap (µm)
1st pass	150	50
2nd pass	40	20
3rd pass	15	10
4th pass	5	5

2.3. Preparation of Composite Films

The as-prepared dispersion was mixed with 1% crosslinker in relation to the weight of LSR and further diluted with toluene to ensure appropriate viscosity. This step was performed in a 3-roll mill at a gap of 30 µm. The dispersion was then subjected to degassing at a low pressure of 0.8 bar for 30 min.

Composite film was made from this degassed dispersion through a spread coating process. In this process, a polymethyl pentene-coated transfer paper (Schöller, Weißenborn, Germany) was stretched between two thin rubberised rollers in a frame supported by springs. The dispersion was applied to the transfer paper in front of the doctor blade and spread into a thin layer with the help of this blade. The space between the paper and the blade was adjusted with the help of a feeler gauge. A wet film of a defined thickness was formed and then dried at 70 °C for 2 min in an oven to remove remaining solvent. A second coating was then made over the first in order to attain a final film of the desired thickness. It was difficult to attain a film thickness between 150 and 200 µm with only a single coat, as this hindered the escape of the air/vapour bubble from the wet film during the drying stage and created a pinhole when the bubble popped during the curing stage. After the second coat, the film was again dried at 70 °C for 2 min to remove the solvent. Subsequently, the dried film was then cured at 180 °C for 6 min. After removing the composite film from the transfer paper, the thickness was measured. Films with thickness between 150 and 200 µm were prepared.

2.4. Application of Contacting Electrodes

In order to apply electrical power to the composite film, the film needed to be electrically contacted by means of a highly conductive coating to ensure reduced power loss at the point of contact. The highly conductive coating was realised using a highly filled silicone dispersion based on silver microparticles. Silver microparticles were mixed with silicone in the presence of a coupling agent. A small amount of toluene was added to aid easy mixing and provide a paste like consistency so that it could be sieve-printed over the composite films. A template made of Teflon (4 mm × 300 mm) was used for printing. Approximately 3 g of the paste was spread out over the template by means of a small screen printing squeegee. Two electrodes were printed on the composite surface at a distance of 5 cm. The printed films were subsequently dried at 100 °C for 2 min to remove the solvents and then cured at 160 °C for 4 min.

2.5. Morphological Characterisation

Cross-sections of the samples were made with a sharp razor blade. The cross-sections were then coated in a very thin layer (3 nm) of gold and observed using a scanning electron microscope, Quanta 250 FEG (FEI, Dresden, Germany). A secondary electron detector (SE2) was used for imaging at an operating voltage of 10 kV with a working distance of approximately 10 mm.

2.6. Electrical Conductivity Measurements

Electrical conductivity was measured with Loresta equipment (Mitsubishi Chemical Analytech Co., Ltd, Japan) using a 4-point method, where 4 electrodes were placed over the composite film. A defined current was applied through the outer electrodes, and voltage drop was measured by the inner electrodes. The conductivity was measured on a composite film (mentioned in Section 2.3) approximately 30 cm × 30 cm in size, with a thickness of between 150 and 200 µm. The conductivity was measured after conditioning the samples for 24 h at 23 °C and 50% relative humidity.

2.7. Heating Behaviour Measurements

The electrically contacted films were cut to a size of 10 cm × 10 cm in which the contacted area was 5 cm × 10 cm. The sample was supported over a stretchable frame by clamping it in rubberised rollers so that the sample did not rest over the table, and there was an air gap in between to avoid conductive heat loss. Electrical power was supplied through a multimeter, with the help of which voltages between 6 and 60 V were realised in steps of 6 or 12 V. Simultaneously, the current flowing through the sample was measured in the multimeter. The temperature distribution over the sample was measured using an IR camera and IRIS plus software. The mean of the temperature distribution was then used for further analysis.

2.8. Heating Behaviour Measurements While Stretching

The sample as mentioned in Section 2.7 was also used here. One of the rubberised rollers could be moved, so that the sample stretched (Figure 1). As it was stretched, the rubberised rollers could be clamped using a screw. In this manner, the samples were stretched by 5%, 10%, 15% and 20% of their initial dimension. After stretching, the same procedure was followed as mentioned in Section 2.6 to measure the heating behaviour at every stage of stretching. After each stage of stretching, the sample was released and allowed to relax for at least 2 h.

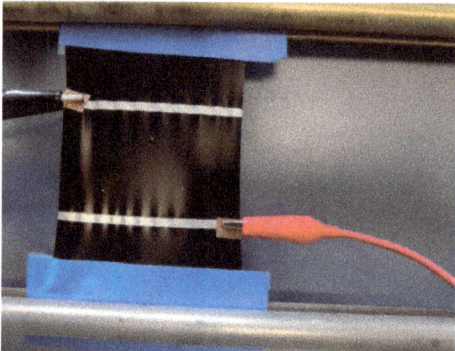

Figure 1. Sample mounted on a stretchable frame.

3. Results and Discussion

3.1. Morphology of Silicone Composites

The morphology of the silicone composites was investigated by SEM observation. In Figure 2, it can be observed that the CNT aggregates were distributed homogeneously throughout the samples, irrespective of the CNT concentration. This was analogous to the observations of Chu et al. [23]. CNTs are purchased in large agglomerates. After 3-roll milling in the presence of polymer, the agglomerates break down into small aggregates, which are responsible for the formation of a conductive network. Figure 2 (top) shows silicone composites with 1% CNT, a relatively low content of CNT as evidenced by the lower quantity of white spots (CNT) on the cross-section. In the case of composites with 5% CNT (bottom), the increase in the number of CNT aggregates is clearly visible. Although the CNT loading was higher, no large agglomerates were observed. This indicated that the CNTs interacted well with the polymer matrix.

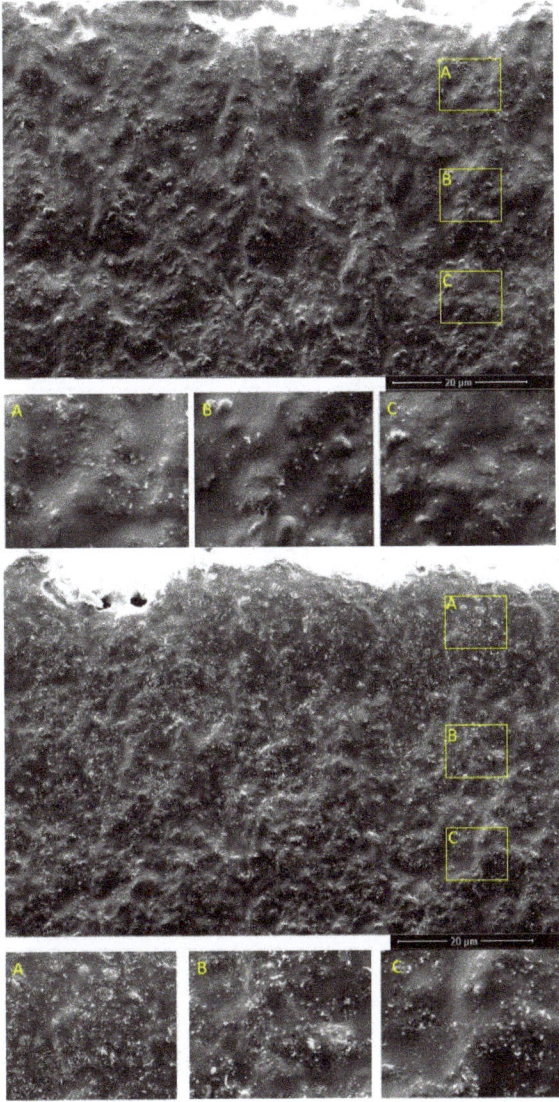

Figure 2. SEM images of silicone composites with 1% CNT (top) and 5% CNT (bottom).

3.2. Electrical Property of Silicone Composites

The electrical conductivities of the silicone composite films are presented below in Figure 3.

The blue curve represents the electrical conductivity of the composites that were supported by substrate, while the orange curve represents the electrical conductivity of the composites that were delaminated from the substrate. As expected, the electrical conductivity of the composites increased with CNT loading. As the CNTs were added to the silicone matrix, at a loading of 1%, the CNT aggregates came into contact with each other, forming a continuous conductive network, allowing electron flow throughout the sample. It was not possible to measure the electrical conductivity of composites with CNT loading of 0.1% and 0.5%, as the conductivity was very low, indicating that percolation occurred at approximately 1% CNT, which is also seen in other works [24,25]. As the CNT content was further raised, the conductive network became denser and provided more pathways for the electrons to flow, thereby leading to an increase in electrical conductivity. It was also observed that while raising CNT content to 3% led to a substantial increase in electrical conductivity, but when the CNT loading further rose to 5%, the upsurge in electrical conductivity was not high. Upon further increase to 7%, conductivity increased by only a marginal increment, yielding a conductivity of 106 S/m. Chu et al. also attained similar values with silicone/MWCNT composites [26,27]. It was observed that the conductivity of the composites did not remain the same after delamination. When the composite was prepared on the substrate, the polymer molecules were in a stressed state. Upon delamination, the molecules tended to move towards an entropically favourable configuration, a coiled state. As the polymer coiled, the CNT aggregates were pushed further from each other, leading to a decrease in the contribution from CNT non-ohmic contacts (nanotubes in the contact were separated by several polymer chains) in the conductive network. As there was only a slight decrease in conductivity, it was inferred that there was no decrease in contribution from ohmic contacts (direct contacts between the nanotubes) [28]. Such a decrease in conductivity on delamination was observed at all concentrations but was more evident at very low concentrations (<2%). Composites ideal for heating applications have a conductivity greater than 10 S/m, limited by considerations concerning safety during operation [29].

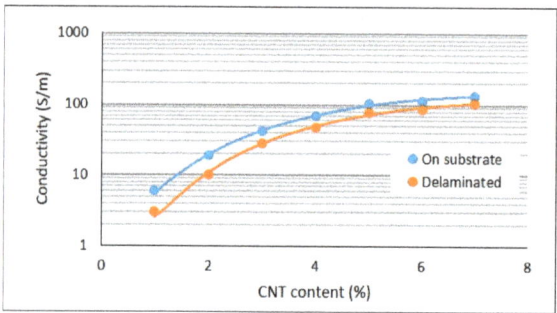

Figure 3. Electrical conductivity of silicone/CNT composites in relation to filler content.

Trends observed in electrical conductivity in relation to strain for silicone composites with varying CNT contents are presented in Figure 4 at four different voltages. It was initially thought that 1% CNT composite would lose its conductivity rapidly when stretched. Contrary to the expectations, for 1% CNT composites, no relationship was observed between conductivity and level of strain. There was no significant loss of density in the conductive network until stretching reached 20%. A similar phenomenon was observed for 3% CNT composites, with only a slight variation. Such behaviour could be due to a greater length of CNT (1.5 μm), while composites with shorter CNTs (0.8 μm) lost their conductivity even at low strains [30]. In the case of the 5% CNT composite, conductivity remained almost constant up to a strain of 5% and then began decreasing considerably. It can be inferred that the conductive network starts disintegrating from a strain of 10%. A similar phenomenon was observed with the 7% CNT composite, wherein conductivity remained constant up to a strain of 5%, and then decreased rapidly. The effect of voltage on the conductivity of composites with lower CNT content was negligible, but at higher CNT contents it was pronounced, especially at 7% CNT. Generally, at higher voltages, more charge carriers are created, which should result in higher conductivity. However, in lower CNT content composites, the number of CNT contacts was very low, acting as a bottleneck for electron flow. Conversely, for composites with higher CNT contents there were numerous CNT contacts; hence, the charge carriers were not interrupted and contributed to the increase in conductivity.

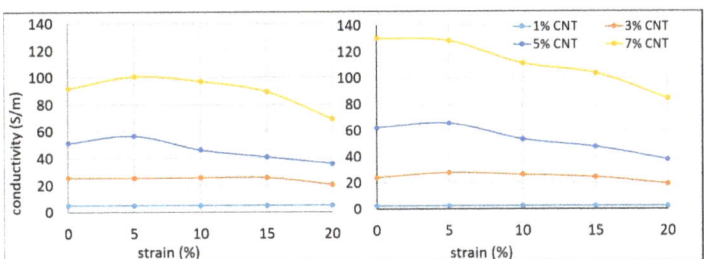

Figure 4. Effect of strain on the conductivity of silicone composites at different CNT concentrations at 6 V (**left**) and 24 V (**right**).

3.3. Heating Behaviour

The heating behaviour is shown in Figure 5. As electrons flow through the CNT, heat is developed as a consequence of the Joule effect.

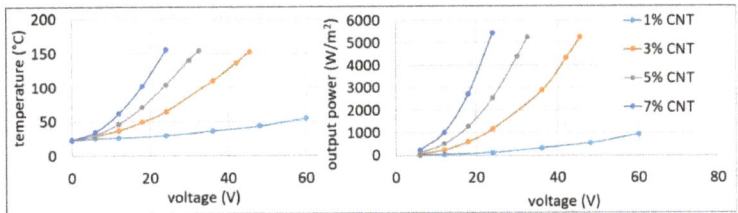

Figure 5. Increase in temperature (**left**) and output power (**right**) with voltage at different CNT concentrations.

Composites with a CNT content of between 1% and 7% were investigated. An example of how the temperature was measured is shown in Figure 6. At 1% loading, the CNTs were already percolated; hence, the composite experienced joule heating upon the application of electrical potential. It was observed that at 1% CNT content, the temperature increase was almost linear with the increase in voltage. The maximum temperature achievable for the 1% CNT composite, which had a conductivity of 3 S/m, was 55 °C at 60 V. There was a marked difference in the heating behaviour for 3% CNT compared to 1% CNT. This was due to almost an order of increase in conductivity (28 S/m) with respect to the 1% CNT composite [31]. The 3% CNT composite reached a temperature of 150 °C at a relatively low voltage (45 V). This temperature was approximately four times higher than that for the 1% composite. It was also able to attain higher temperatures, but to avoid thermal degradation of the composite, the maximum temperature was limited to below 160 °C. Moreover, in contrast to the linear correspondence of temperature to voltage in the case of the 1% composite, an exponential increase was observed. In the case of the 5% CNT composite, the temperature rise was rapid and reached a temperature of 155 °C at a lower voltage of 33 V. In the case of the 7% CNT composite, with an electrical conductivity of 106 S/m, a steep increase in temperature was observed, rising to 155 °C at a very low voltage of 23 V.

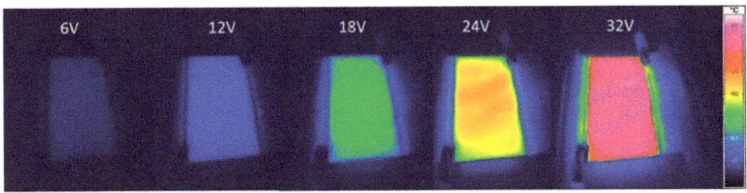

Figure 6. IR images of silicone/5% CNT composite with 5 cm electrode distance at different potential.

3.4. Heating Behaviour When Uniaxially Strained

The heating behaviour of silicone composites was investigated when the samples were strained, as this relationship is relevant in most heating applications. The heating behaviour of the 1% CNT composite was represented in two formats. The plots on the left show the variations with voltage and the plots on the right with strain.

For each sample, the composite was stretched to four different strain levels (5%, 10%, 15%, and 20%) in a direction perpendicular to the electrodes. As the sample was elongated, the connections between CNTs were reduced, and the conductive network became loose, resulting in a fall-off in conductivity. In Figure 7, it can be observed that the heating behaviour of the composite up to 24 V was almost constant at all the strain levels. The same applies to the observations of Figure 4, which states that the conductivity of the 1% CNT composite remained almost constant at all strain levels up to a potential of 24 V. Above 24 V, heating behaviour when strained up to 10% was almost the same, with very slight differences at higher voltages. However, heating behaviour at 20% strain was considerably lower than at other strain levels. While at 10% strain or less, there was little change in heating behaviour, at 20% strain, there was a considerable difference, which can be attributed to the fact that the CNTs had moved apart considerably and the number of contacts were reduced, leading to fewer pathways for electron flow.

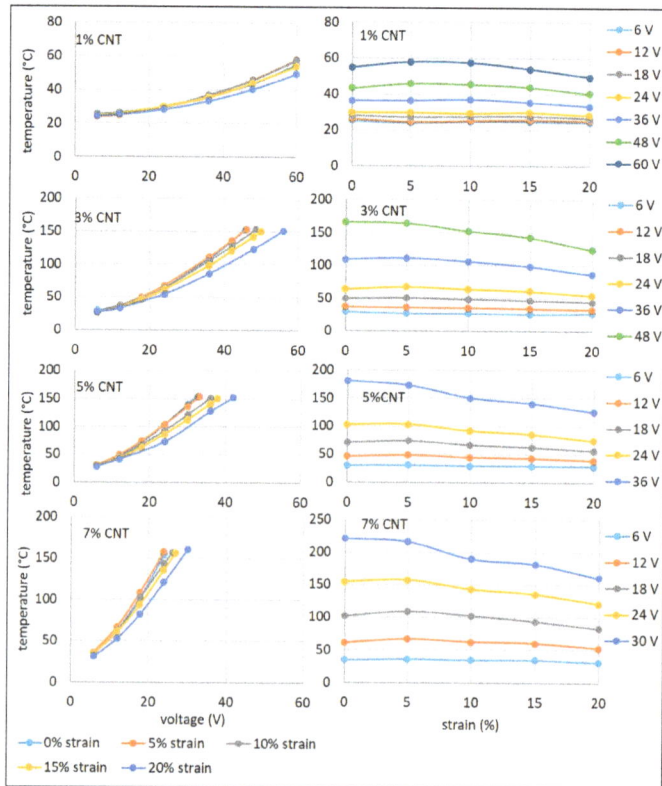

Figure 7. Increase in temperature with voltage (**left**) and with strain (**right**) at different CNT contents.

In comparison to the 1% CNT composite, the temperatures achievable at different strains were much higher for the 3% CNT composite (Figure 7). This was primarily due to the denser conductive network being affected by the higher CNT loading, thereby creating more pathways for electrons to produce Joule heating [32]. As observed for the 1% CNT composite, heating behaviour remained the same up to 24 V at all strain levels. However, at higher potential, though higher temperatures were reached, when the composite was strained to 10%, the temperature started to deteriorate; at 15% strain, there was further decrease; and at 20% strain, there was a significant difference.

In comparison to the 3% CNT composite, higher temperatures were achieved at much lower voltages for the 5% CNT composite at all strain levels (Figure 7). As observed for the 3% CNT composite, heating behaviour remained the same up to 12 V at all strain levels. However, at 18 V, there was a slight decrease in temperatures at higher strains. At 24 V, the temperature remained constant up to a strain of 5%; when strained to 10%, the temperature started to deteriorate; at 15% strain, there was a further decrease; and at 20% strain, there was a considerable difference. At 36 V, the temperature started to decrease immediately from a strain of 5% and then decreased extensively at higher levels of strain.

In comparison to the 5% CNT composite, higher temperatures were achieved at relatively low voltages for the 7% CNT composite at all strain levels (Figure 7). As observed for the 3% CNT composite, the heating behaviour remained the same up to 12 V at all strain levels. However, at 18 V the temperature was almost constant up to 10% strain, and then there was slight decrease in temperature at higher strains. At higher voltages (24 and 30 V) up to a level of 5% strain, the temperature was constant, and then it gradually decreased.

4. Conclusions

The delicate but efficient mixing method using a 3-roll mill yielded silicone/CNT composites up to a loading of 7% CNT, with good filler dispersion even at such a high content. Good dispersion quality was reflected in the electrical conductivity achieved, with a conductivity of 106 S/m attained at 7% CNT content. Interestingly, composites with lower filler content (1% and 3%) exhibited conductivity independent of strain level owing to the lower density of the conductive network. On the contrary, composites with higher filler content (5% and 7%) displayed strong strain-dependent conductivity, especially at strain levels above 10%. Such behaviour can be attributed to the loosening of the denser conductive network. The sharp rise in temperature for composites with higher CNT content was by virtue of their higher conductivity, whereby a 7% CNT composite was able to achieve a temperature of 155 °C at a lower voltage of 23 V. In the case of composites containing lower CNT content (1% and 3%), only at very high voltages (from 36 V) did strain level affect the rise in temperature. Conversely, for composites containing higher CNT content (5% and 7%), the influence of strain level started at lower voltages (from 18 V). Such behaviour can again be attributed to the density of the conductive network, in which at lower filler contents, the change in conductive network structure is lesser than at higher filler contents. For applications demanding very little temperature change, it is advisable to select composites with lower CNT content, while for applications that demand lower input voltage, composites with higher CNT content would be the wiser choice.

Author Contributions: Conceptualisation, K.T., M.G. (Maik Gude) and A.B.; methodology, M.G. (Minoj Gnanaseelan); software, M.G. (Minoj Gnanaseelan); validation, M.G. (Minoj Gnanaseelan), R.S., R.K., and B.P.; writing—original draft preparation, M.G. (Minoj Gnanaseelan), B.P.; writing—review and editing, R.S. and B.P.; visualisation, M.G. (Minoj Gnanaseelan), R.S. and B.P.; supervision, K.T, M.G. (Maik Gude), R.K., and A.B.; project administration, K.T., M.G. (Maik Gude), and A.B.; funding acquisition, K.T., M.G. (Maik Gude), and A.B. All authors have read and agreed to the published version of the manuscript.

Funding: This research (grant number 266 EBR) was funded by BMWi—Federal Ministry of Economics and Technology in Germany and by National Center for Research and Development in Poland (grant number CORNET/27/2/2020). The authors extend their gratitude for the financial aid provided.

Data Availability Statement: The data presented in this study are available on request from the corresponding author.

Conflicts of Interest: The authors declare no conflict of interest.

References

1. Mittal, V. Polymer layered silicate nanocomposites: A review. *Materials* **2009**, *2*, 992–1057. [CrossRef]
2. Yu, Y.-Y.; Chen, C.-Y.; Chen, W.-C. Synthesis and characterization of organic–inorganic hybrid thin films from poly (acrylic) and monodispersed colloidal silica. *Polymer* **2003**, *44*, 593–601. [CrossRef]
3. Kim, I.-H.; Jeong, Y.G. Polylactide/exfoliated graphite nanocomposites with enhanced thermal stability, mechanical modulus, and electrical conductivity. *J. Polym. Sci. Part B Polym. Phys.* **2010**, *48*, 850–858. [CrossRef]
4. Shit, S.C.; Shah, P.M. A review on silicone rubber. *Natl. Acad. Sci. Lett.* **2013**, *36*, 355–365. [CrossRef]
5. Rahimi, A.; Mashak, A. Review on rubbers in medicine: Natural, silicone and polyurethane rubbers. *Plast. Rubber Compos.* **2013**, *42*, 223–230. [CrossRef]
6. Hamdani, S.; Longuet, C.; Perrin, D.; Lopez-Cuesta, J.M.; Ganachaud, F. Flame retardancy of silicone-based materials. *Polym. Degrad. Stab.* **2009**, *94*, 465–495. [CrossRef]
7. Lucas, P.; Robin, J.-J. Silicone-based polymer blends: An overview of the materials and processes. *Funct. Mater. Biomater.* **2007**, *209*, 111–147.
8. Yoshimura, K.; Nakano, K.; Hishikawa, Y. Flexible tactile sensor materials based on carbon microcoil/silicone-rubber porous composites. *Compos. Sci. Technol.* **2016**, *123*, 241–249. [CrossRef]
9. Varga, Z.; Filipcsei, G.; Zrínyi, M. Magnetic field sensitive functional elastomers with tuneable elastic modulus. *Polymer* **2006**, *47*, 227–233. [CrossRef]
10. Park, E.-S. Mechanical properties and processibilty of glass-fiber-, wollastonite-, and fluoro-rubber-reinforced silicone rubber composites. *J. Appl. Polym. Sci.* **2007**, *105*, 460–468. [CrossRef]
11. Chen, L.; Lu, L.; Wu, D.; Chen, G. Silicone rubber/graphite nanosheet electrically conducting nanocomposite with a low percolation threshold. *Polym. Compos.* **2007**, *28*, 493–498. [CrossRef]
12. Ding, T.; Wang, L.; Wang, P. Changes in electrical resistance of carbon-black-filled silicone rubber composite during compression. *J. Polym. Sci. Part B Polym. Phys.* **2007**, *45*, 2700–2706. [CrossRef]
13. Kim, H.I.; Wang, M.; Lee, S.K.; Kang, J.; Nam, J.D.; Ci, L.; Suhr, J. Tensile properties of millimeter-long multi-walled carbon nanotubes. *Sci. Rep.* **2017**, *7*, 9512. [CrossRef] [PubMed]
14. Vast, L.; Mekhalif, Z.; Fonseca, A.; Nagy, J.B.; Delhalle, J. Preparation and electrical characterization of a silicone elastomer composite charged with multi-wall carbon nanotubes functionalized with 7-octenyltrichlorosilane. *Compos. Sci. Technol.* **2007**, *67*, 880–889. [CrossRef]
15. Witt, N.; Tang, Y.; Ye, L.; Fang, L. Silicone rubber nanocomposites containing a small amount of hybrid fillers with enhanced electrical sensitivity. *Mater. Des.* **2013**, *45*, 548–554. [CrossRef]
16. Shi, X.; Jiang, B.; Wang, J.; Yang, Y. Influence of wall number and surface functionalization of carbon nanotubes on their antioxidant behavior in high density polyethylene. *Carbon* **2012**, *50*, 1005–1013. [CrossRef]
17. Yamane, S.; Ata, S.; Chen, L.; Sato, H.; Yamada, T.; Hata, K.; Mizukado, J. Experimental analysis of stabilizing effects of carbon nanotubes (CNTs) on thermal oxidation of poly (ethylene glycol)–CNT composites. *Chem. Phys. Lett.* **2017**, *670*, 32–36. [CrossRef]
18. Ata, S.; Hayashi, Y.; Thi, T.B.N.; Tomonoh, S.; Kawauchi, S.; Yamada, T.; Hata, K. Improving thermal durability and mechanical properties of poly (ether ether ketone) with single-walled carbon nanotubes. *Polymer* **2019**, *176*, 60–65. [CrossRef]
19. Moniruzzaman, M.; Winey, K.I. Polymer nanocomposites containing carbon nanotubes. *Macromolecules* **2006**, *39*, 5194–5205. [CrossRef]
20. Bauhofer, W.; Kovacs, J.Z. A review and analysis of electrical percolation in carbon nanotube polymer composites. *Compos. Sci. Technol.* **2009**, *69*, 1486–1498. [CrossRef]
21. Bokobza, L. Enhanced electrical and mechanical properties of multiwall carbon nanotube rubber composites. *Polym. Adv. Technol.* **2012**, *23*, 1543–1549. [CrossRef]
22. Wang, L.; Wang, X.; Li, Y. Relation between repeated uniaxial compressive pressure and electrical resistance of carbon nanotube filled silicone rubber composite. *Compos. Part A Appl. Sci. Manuf.* **2012**, *43*, 268–274. [CrossRef]
23. Chu, K.; Park, S.-H. Electrical heating behavior of flexible carbon nanotube composites with different aspect ratios. *J. Ind. Eng. Chem.* **2016**, *35*, 195–198. [CrossRef]
24. Chu, K.; Yun, D.-J.; Kim, D.; Park, H.; Park, S.-H. Study of electric heating effects on carbon nanotube polymer composites. *Org. Electron.* **2014**, *15*, 2734–2741. [CrossRef]
25. Yan, J.; Jeong, Y.G. Synergistic effect of hybrid carbon fillers on electric heating behavior of flexible polydimethylsiloxane-based composite films. *Compos. Sci. Technol.* **2015**, *106*, 134–140. [CrossRef]
26. Chu, K.; Kim, D.; Sohn, Y.; Lee, S.; Moon, C.; Park, S. Electrical and thermal properties of carbon-nanotube composite for flexible electric heating-unit applications. *IEEE Electron Device Lett.* **2013**, *34*, 668–670. [CrossRef]
27. Trommer, K.; Morgenstern, B.; Petzold, C. Preparing of Heatable, CNT-Functionalized Polymer Membranes for Application in Textile Composites. *Mater. Sci. Forum* **2015**, *825*, 67–74. [CrossRef]
28. Kumar, V.; Lee, G.; Choi, J.; Lee, D.-J. Studies on composites based on HTV and RTV silicone rubber and carbon nanotubes for sensors and actuators. *Polymer* **2020**, *190*, 122221. [CrossRef]
29. Kutějová, L.; Vilčáková, J.; Moučka, R.; Natalia, E.; Babayan, V. A solvent dispersion method for the preparation of silicone composites filled with carbon nanotubes. *Chem. Listy* **2014**, *108*, 78–85.

30. Moseenkov, S.I.; Zavorin, A.V.; Ishchenko, A.V.; Serkova, A.N.; Selyutin, A.G.; Kuznetsov, V.L. Using Current-Voltage Characteristics to Control the Structure of Contacts in Polyethylene Based Composites Modified by Multiwalled Carbon Nanotubes. *J. Struct. Chem.* **2020**, *61*, 628–639. [CrossRef]
31. Wang, P.; Peng, Z.; Li, M.; Wang, Y. Stretchable Transparent conductive films from long carbon nanotube metals. *Small* **2018**, *14*, 1802625. [CrossRef] [PubMed]
32. Zeng, Y.; Liu, H.; Chen, J.; Ge, H. Effect of strain on the electrical resistance of carbon nanotube/silicone rubber composites. *J. Wuhan Univ. Technol. Mater. Sci.* **2011**, *26*, 812–816. [CrossRef]

MDPI
St. Alban-Anlage 66
4052 Basel
Switzerland
Tel. +41 61 683 77 34
Fax +41 61 302 89 18
www.mdpi.com

Materials Editorial Office
E-mail: materials@mdpi.com
www.mdpi.com/journal/materials

www.ingramcontent.com/pod-product-compliance
Lightning Source LLC
LaVergne TN
LVHW070553100526
838202LV00012B/456